BIOECONOMIC ANALYSIS OF FISHERIES

Professor R. Hannesson
Norwegian School of Economics and Business Administration, Bergen

Published by arrangement with
the Food and Agriculture Organization of the United Nations
by
Fishing News Books

Fishing News Books
A division of Blackwell Scientific
Publications Ltd
Editorial Offices:
Osney Mead, Oxford OX2 OEL
25 John Street, London WC1N 2BL
23 Ainslie Place, Edinburgh EH3 6AJ
283 Main Street, Cambridge,
MA 02142, USA
54 University Street, Carlton,
Victoria 3053, Australia

Other Editorial Offices:
Librairie Arnette SA
2, rue Casimir-Delavigne
75006 Paris
France

Blackwell Wissenschafts-Verlag
Meinekestrasse 4
D-1000 Berlin 15
Germany

Blackwell MZV
Feldgasse 13
A-1238 Wien
Austria

First published 1993

Set by Best-set Typesetter Ltd, Hong Kong

DISTRIBUTORS

Marston Book Services Ltd
PO Box 87
Oxford OX2 0DT
(*Orders*: Tel: 0865 240201
Fax: 0865 721205
Telex: 83355 MEDBOK G)

USA
Blackwell Scientific Publications Inc
283 Main Street
Cambridge, MA 02142
(*Orders*: Tel: 800 759-6102)

Canada
Oxford University Press
70 Wynford Drive
Don Mills
Ontario M3C 1J9
(*Orders*: Tel: (416) 441-2941)

Australia
Blackwell Scientific Publications Pty Ltd
54 University Street
Carlton, Victoria 3053
(*Orders*: Tel: (03) 347-5552)

British Library
Cataloguing in Publication Data
A Catalogue record for this book is available
from the British Library

ISBN 0-85238-198-0

Library of Congress Cataloging-in-Publication
Data
Hannesson, Rögnvaldur.
Bioeconomic analysis of fisheries/
R. Hannesson.
p. cm.
Includes bibliographical references (p.)
and index.
ISBN 0-85238-198-0
1. Fisheries – Economic aspects. 2. Fish
populations. 3. Fishery management.
I. Title.
SH334.H35 1992
338.3′727 – dc20 92-26496
CIP

Contents

Preface

Addressing two different readerships at the same time is a risky undertaking. It is like aiming at two sitting ducks at the same time; most likely you'll miss both. Yet it seems worthwhile to attempt to write a book about fisheries management that simultaneously addresses both economists and fisheries biologists, or students of either discipline. Fisheries management is indeed a multidisciplinary subject. Obviously it must be based on sound biological expertise, but the ultimate objective is economic. Fisheries management that disregards economic aspects may succeed in preserving fish stocks, but it will waste other resources such as investment funds and labour, and it is likely to be distinctly unhelpful for the fishing industry.

The way the book addresses the perceived needs of fisheries biologists and economists is as follows. The basic ideas of economics and how they apply to the fishing industry are introduced in a non-rigorous manner intelligible to people without much knowledge of mathematics. This should be useful for both biologists and others without any formal training in economics who have become involved with fisheries management. For the economist the book introduces mathematical models of fisheries, tools which are necessary for economic analysis of the fishing industry.

The contents of the book are as follows. Chapter 1 is an introduction to the main concepts of economic analysis and how they apply to the fishing industry. This is a non-technical discussion comprehensible by the interested layman. Chapter 2 introduces the economic analysis of fisheries and compares economically optimal fishing with unregulated fishing. This discussion is also for the most part non-technical, with the formal analysis being presented in a self-contained section (2.8).

Chapter 3 discusses three popular fishery models developed by fisheries biologists. This should be a useful chapter for economists who need tools for applied analysis of fisheries, but it should also be of interest for fisheries biologists who have been trained in using these models for purely biological analysis but may be unaware of how they can be extended to incorporate economic analysis.

Chapter 4 deals with the problems arising from random fluctuations in fish stocks. Should catches be stabilized despite fluctuations in the stocks, or do we have to accept that catches will fluctuate as well? If so, how does this affect the economics of the fishing industry? These are important questions but are all too often ignored. They are also quite difficult to deal with by formal analysis. This is the most difficult chapter mathematically and presupposes knowledge of differential and integral calculus.

The last chapter deals with the political economy of fishery regulations. Is there any reason to expect that regulated fisheries will do any better, from a social point of view, than unregulated fisheries, despite the well-known shortcomings of the latter? The answer hinges on what the objectives of fisheries managers are. Like the first two chapters, this discussion is informal and intelligible by the interested layman. Readers with a weak background in mathematics might want to skip Chapter 4 and possibly Chapter 3 and go directly to Chapter 5 after having read the first two chapters.

Rögnvaldur Hannesson
Bergen

Chapter 1
Introduction

This book is about bioeconomic analysis of fisheries. The importance of such analysis needs no elaboration. Fish resources are an important source of food, the need for which is steadily growing. The sustained use and existence of fish resources depends on their reproduction, the biology of which is complex and not well understood. The prudent use of fish resources is, however, by no means a matter only of biological data and analysis. Economic analysis deals with the use of scarce resources, of which fish are but one example. It attempts to explain why resources are used in the way they are and how they could possibly be used better. As our needs outstrip our capacity to satisfy them, there are choices to be made with respect to how we use the resources at our disposal. Economic analysis can make such choices more enlightened by specifying the costs and benefits of every possible choice.

The basic economic choice to be made with respect to living resources such as fish stocks is how intensively they should be exploited. Fishing requires the application of manpower, fuel and various implements, all of which could be used for some other purpose. The question we must consider is whether we are getting as much value in return for our efforts when fishing as we would be getting otherwise. Economic analysis of fishing is particularly important because the absence of private ownership of fish resources encourages overexploitation, possibly to the point of irreversible depletion. This can be corrected by regulating the fishing industry, but regulations are not likely to be successful unless enlightened by economic analysis and supported by a structure of incentives encouraging rational exploitation.

1.1 How a market economy works

This subject is big and important enough to require a book of its own. The introduction by Dorfman (1967) is, despite its age, still one of the best. For those readers who are unfamiliar with economics an explanation of some of its basic ideas and how a market economy works will be given. It is customary in economic analysis to distinguish between two types of agents in the economy, producers and consumers. Producers produce goods and services, which in turn are bought by consumers, who use them. To these it is appropriate to add a third type of agents, the regulators. Regulators are politicians and civil servants who regulate the economic process directly or indirectly through setting the rules by which producers and consumers have to play.

Each agent is supposed to act in his or her best interest, a behaviour that for analytical purposes can be formalized as maximization of the utility one derives from one's endeavours. Producers are supposed to maximize profits, because this gives them the greatest possible command over resources. Consumers dispose of their given budgets (including time) in a way that gives them the greatest possible satisfaction. Regulators intervene in the economic system to maximize their own benefit, which for a politician is likely to mean maximizing the likelihood of being elected, and, for a civil servant, maximizing career opportunities. The public will be well served by the endeavours of regulators to the extent they happen to promote the public interest simultaneously with their own.

We shall return to the regulators in the last chapter of this book, but let us for now concentrate on producers and consumers. Producers can do a variety of things to maximize their profits. If a producer is able to manipulate the market price of his product, he may certainly be expected to do so. This is possible, however, only for producers who are responsible for a major portion of the total supply in a given market. Here we shall be concerned mainly with the case where the individual producer is unable to influence the market price.

Having to take prices as being given by the market, one of the most important decisions to be taken in order to maximize profit is how much to produce and to sell, or whether to produce a particular commodity at all. There are, needless to say, many other factors that affect the profits of a business operation; product quality is one example. Even if these are important, it would take us too far to consider them here. What is particularly important for how intensively fish stocks are being utilized is how many firms are engaged in the fishing industry and how much fish each firm produces. This will be a central theme in the discussion to follow.

The idea that consumers try to maximize the utility from a given budget leads to an interesting theory about the information content of prices. Imagine a consumer who carefully plans how to spend his money. He has allocated the entire budget down to the last dollar (or whatever the currency unit is). There will be a wide range of choices with regard to how to spend this last dollar. Now, if the budget has been optimally allocated among all possible items, it must be a matter of indifference how the last dollar is spent. Hence the last dollar must provide the same utility, regardless of how it is spent. A product that costs twice as much as another product must therefore, from the point of view of the consumer, provide twice as great utility.

Note that this equivalence of utility and price of a product refers to choices 'at the margin', that is, the last unit of a commodity or service bought, not the entire amount. The marginal utility and the price of commodities are related to their relative scarcity, not to intrinsic values. Consider the old paradox of value. Why is it that diamonds are expensive and water is cheap (in most climates), even if we can live perfectly well without diamonds but not at all without water? The answer lies in their relative scarcity; diamonds are hard to come by while there is usually plenty of water. If it were the other way around, this would be reflected in their prices. There are indeed places on earth where water is scarce and where its price is accordingly high.

This information content of prices becomes important when we consider how producers react to prices. A high price of a commodity makes it attractive to increase the supply of that commodity, other things being equal. But how far should the supply of the commodity be increased? As long as additional production of the commodity increases the profit of the producer he will be interested in expanding the production. But eventually a point is reached at which one additional unit supplied would increase the costs of production by an amount equal to its price, and the incentive to increase production disappears.

This is exactly the point where the increase in supply should stop, provided that the marginal cost of production (the cost of an additional unit) reflects the value of the commodities people must go without in order to increase the supply of the commodity in question. Increasing the supply of a given commodity means that the supply of other commodities must be less than otherwise, unless some productive resources are idle. In order to increase the supply of a given commodity one needs manpower, machines, buildings, tools, fuel, etc. All these 'inputs', or 'factors of production', as they are often called, could be used to produce something else, be it agricultural products or paved roads. The value of this forgone production is the real cost of providing a particular commodity; the utility of consumers is reduced by approximately the price of each commodity or service times the quantity forgone. For the producer the marginal cost of a commodity is equal to the cost of the inputs required to produce an additional unit. In a perfect market system where the prices of all inputs reflects the value they could produce if used otherwise, the marginal cost of commodities is equal to the value of commodities that must be forgone in order to increase the supply of a given commodity by one unit.

So, to return to the water and diamond example, the reason why the supply of diamonds is not increased to the point where their price would be equal to the price of water is the high cost of providing diamonds; the value of goods and services that must be given up in order to provide an additional amount of diamonds is much greater than needed to provide an additional amount of water. This is a solution that producers attain on the basis of their attempt to maximize profits. This behaviour has, perhaps surprisingly, a great potential to benefit consumers. As long as the price of a commodity is greater than its marginal cost, producers will be motivated to increase the supply of that commodity. This is in the interest of the consumer; the value of the goods that must be given up in order to increase the supply of the commodity in question by one unit is less than what that unit is worth to consumers.

1.2 Market failure

The rosy picture painted above is almost too good to be true. Those who wish to maintain that the world is characterized by a high degree of harmony of economic interests have a hard time explaining class conflicts, strikes, rows over international trade, etc. Yet there is more than a grain of truth in the reasoning above; the market system has a great potential to ensure an efficient use of resources, to the benefit of all. What the market system is not particularly good at is distributing incomes in an equitable manner. There are other devices (taxes and transfers, for example) available to deal with distributional issues, which are beyond the scope of this book. Suffice it to say that for any distribution of incomes, be it good or bad, the market system is often successful at ensuring an efficient use of resources, but sometimes it fails to do so. There are many reasons for such failures. One such reason, which is particularly relevant for the fishing industry, is the absence of ownership of scarce resources; or, what amounts to the same thing, the fact that scarce resources can be used without any cost, as if they were unlimited. One of the critical conditions that must be satisfied if the market system is to work well is that all scarce resources carry a price that correctly reflects their scarcity.

Later we shall discuss this 'market failure' in the context of fisheries, but let us briefly sketch the argument here. Fish are a scarce resource; there is not an unlimited amount of fish for us to take. Fish resources do replenish themselves through natural growth but they are nevertheless scarce; their growth is affected by how much we take, and they could possibly be eliminated altogether if too much were taken in any given period. Yet the fishing industry is seldom forced to pay anything for the fish it takes, apart from the necessary costs of going out to sea to get it. Not surprisingly, this leads to overexploitation of the fish resources. This is the core of the argument why fisheries management is needed.

It may be added that fish are by no means unique in this regard; the same is generally true of environmental resources. Industrial firms have in the past to a large extent been able to dispose of their waste products to the environment (air, water, soil) free of charge. This had led to excessive pollution which now is being ever more tightly regulated. Such regulation is being increasingly pursued by economic incentives of a kind similar to those discussed in the final chapter of this book.

Chapter 2
The Economic Theory of Fisheries

2.1 Surplus growth and natural equilibrium

It is natural to begin a bioeconomic analysis of fisheries by recognizing the biological limits to fishing. We shall start from some basic, stylized facts. Many fish stocks have been utilized from time immemorial and thus proved capable of withstanding a sustained exploitation, provided that the limits of self-renewal are not exceeded. A reasonable starting point is, therefore, to assume that unexploited fish stocks maintain a natural equilibrium in which growth and natural decay balance, while exploited stocks generate surplus growth which compensates for the amount being fished. This leads to a stock–growth relationship of the kind shown in Fig. 2.1. The curve $G(S)$ in Fig. 2.1 shows the rate of surplus growth per unit of time, for example, tonnes per year (mathematically we are thinking of an instantaneous rate; that is, growth over a 'short' period of time divided by the length of the time period).

Suppose we start with an unexploited stock in a natural equilibrium at point S^* in Fig. 2.1. Initially growth and decay will no longer balance; fishing will cause an additional decay, and the stock will diminish. Ultimately the stock would be depleted, unless the surplus growth rate increases as the stock becomes smaller, as shown by the right half of the growth curve in Fig. 2.1.

Obviously there must be some fish present to generate surplus growth; if the stock has been totally depleted, the growth cannot possibly be positive. Hence the growth rate must reach a maximum at a stock level somewhere between zero and the natural equilibrium level and then fall towards zero. The surplus growth rate could quite possibly become zero and turn negative well before the stock has been totally depleted. In that case there would be a minimum viable stock level, S^{**}, at which the surplus growth rate becomes zero. A minimum level of viability that is significantly greater than zero has important implications for the development and management of the fishery. These will be discussed later in this chapter.

The stock–growth relationship that we have just sketched is a highly simplified picture based on common sense, but useful for understanding the basic mechanics of fishery development. There are various reasons why a reduction of the stock generates surplus growth. Each fish may get a greater share of a given food supply

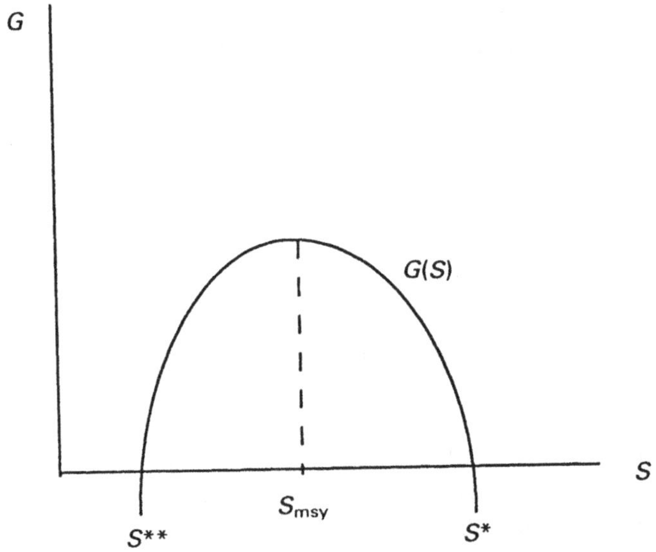

Fig. 2.1 Relationship between stock size (S) and surplus growth (G) per unit of time.

as the number of fish in the stock diminishes. This is likely to increase the growth rate of the fish. Another reason is that exploitation of fish populations changes their age structure towards a greater predominance of young, fast-growing individuals. A third possible reason is that smaller spawning stocks and fewer eggs being fertilized may mean less competition for food and greater survival at a critical stage. In reality all of these are likely to play a role.

Even if the stock–growth relationship is important, it must be emphasized that the growth rate of fish stocks depends on other factors besides the size of the stock itself, such as the supply of food, the predation on the stock by other species, ocean currents and temperature, and possibly more. The proposed relationship between the rate of growth and the stock size may be expected to hold in the long term and on the average, while the actual rate of growth will vary over time for any given level of the stock. The concept of natural equilibrium proposed above should therefore be thought of as a level around which the stock will fluctuate, possibly over a wide range, if it is left unexploited. These fluctuations are certainly important and may pose formidable challenges to fisheries development and management, but the proposed 'deterministic' concepts are useful nevertheless for developing a basic understanding of fisheries.

2.2 The concept of fishing effort

Another fundamental step in the bioeconomic analysis of fisheries is to develop a relationship between the activities of a fishing fleet and the catch it takes. This is not an easy task, for a number of reasons. Fishing fleets typically consist of various types of vessels, equipped in different ways and employing crews with different skills. This is partly the result of technological progress making old techniques obsolete as far as renewal of vessels is concerned, while vessels and equipment that are obsolete in the long term can still be profitably used before they are worn out. The diversity of fishing fleets could, however, be the result of different judgments or preferences among individual decision makers; the type of fishing boat which appears most profitable to one particular fisherman or fishing firm might not be seen that way by all other fishermen or fishing firms. We shall not deal with these problems here but assume that the size and activity of the fishing fleet can be expressed in units of identical, representative vessels.

This brings us to the concept of fishing effort. This concept was originally developed by fisheries biologists for the purpose of fish stock assessment. The fisheries biologists tried to find a measure of fishing activity that would be directly proportional to the mortality it generates in the stock, that is, how large a fraction of the stock it removes per unit of time. Provided that such a measure can be found, the catch per unit of effort will be directly proportional to the size of the exploited stock and can thus be used as an index of the size of the fish stock.

There are, however, two problems with this. The first is that the mortality in a fish stock caused by activity which is directly aimed at removing fish from the stock is also related to the spatial distribution of the stock. By definition, the mortality of a stock is the number of fish removed from the stock per unit of time. Consider now a stock that is always uniformly distributed over a given area. An application of a unit of effort (say, so many hours of trawling) would remove a certain fraction of the fish in the stock. The assumption of a uniform distribution implies that the fish would redistribute themselves instantaneously over the same area. Again applying the same unit of effort would remove the same fraction of the stock as before, but fewer fish since there are fewer fish to be taken. The mortality generated would be proportional to the amount of effort, and the catch per unit of effort would reflect the number of fish in the stock.

But what if the fish react differently to the depletion caused by fishing and keep their density constant, so that the area over which they are spread decreases proportionately to their number? Applying a unit of effort in the area where the fish are located would then always remove the same number of fish, but the mortality generated by a unit of effort would be greater the fewer fish there are in the stock.

These two cases are fairly stylized but nevertheless widely believed to come close to being accurate descriptions of some real world fisheries. The first case is generally believed to be a good approximation for bottom dwelling fish like cod. There is some evidence to support this (Hannesson 1983, Flåten 1987). The second case is widely believed to be a good approximation for fish that live close to the surface and move around in large shoals. If the stocks are depleted there will be fewer shoals, but the size of the shoals is not much affected. Again there is some evidence to support this (Bjørndal 1987, Butterworth 1981, Ulltang 1980).

Another problem with this definition of effort is its narrowness of scope. It considers only that part of the activity of a fishing fleet which is directly aimed at removing fish from the sea. But there are other activities involved in getting a marketable catch of fish. The fishing vessels have to steam between the fishing grounds and the port of landing, the catch and the fishing gear need some handling at sea, and it may be necessary to search for suitable concentrations of fish. All these activities require the use of inputs such as fuel, equipment, manpower, etc. From an economic point of view, all use of such inputs must be taken into account. If these 'auxiliary' activities were always proportionate to fishing effort in the narrow sense (the activity directly aimed at catching the fish) this would not cause any problem, but there is no presumption that such is the case. As fish stocks are depleted and the catch per unit of effort decreases, it is likely that fishing will take more time compared with handling the gear and the catch and sailing to and from port, while searching for fish is likely to take longer.

It seems impossible to derive any generally valid relationship between fishing effort in the narrow sense and fishing effort in the economic sense. This relationship will depend on the characteristics of each fishery, such as whether or not vessels are equipped for long trips or not. We shall not, therefore, pursue this matter further here, but base our further reasoning on the narrow definition of effort, implicitly assuming that the 'auxiliary' activities are proportionate to the fishing effort in the narrow sense. We shall, however, define fishing effort more precisely as the mortality generated in a stock of a given size and distribution. This allows us to take account of the possibility that the distribution of the fish may change in various ways as the stock becomes depleted. Above we have looked at what appear to be extreme cases, one in which a stock of fish is always uniformly distributed over a given area, and one in which its density is constant while the area it occupies shrinks in proportion to the size of the stock. In the exposition to follow we shall make extensive use of these two different cases, as they are particularly easy to handle mathematically. It should be kept in mind, however, that reality is likely to lie somewhere in between, with certain stocks coming close to one or the other.

2.3 Revenue, cost, profit and rent

To analyse fisheries from an economic point of view we need to develop some basic economic concepts, such as revenue, cost, profit and rent. *Revenue* is the price of a product times the quantity sold. A single seller who contributes only a small fraction of the total supply of fish in a given market cannot influence the price to any appreciable extent by varying the amount sold; for each agent in a market supplied by many 'small' suppliers the price is an outcome of the interaction among many agents, on which each has little or no influence. From the aggregate point of view this is not necessarily so; the price of, say, fish may vary greatly with the amount of fish bought and sold within the relevant period (day, week, month or whatever). This will have to be taken into account when considering the aggregate revenue of the fishery.

The concept of *cost* is somewhat more difficult. One basic distinction to be made is between long-term and short-term costs. *Short-term costs*, or *variable costs*, are such cost items that can be varied at short notice. Fuel cost is probably the best example of this; whether or not to buy fuel, and how much, typically is a decision that can be taken at short notice, daily or weekly, depending on the reliability of supplies. Labour costs may be of this kind as well; it may be possible to hire labour even on a daily basis, but typically one has to provide employment for a longer period (paying labour a share of the catch value makes labour cost a short-term cost). The cost of a boat on the other hand is a typical *long-term cost*, or *fixed cost*. The boat has to be paid for when acquired, typically with a loan which must be paid back in instalments and on which one must pay interest, regardless of whether the boat is being used or not.

Another distinction between different types of cost is the social opportunity cost versus private cost. *Private costs* are such expenditures incurred by private individuals or firms as were just mentioned. As explained in Chapter 1, the *social opportunity cost* of providing fish is the value of the goods that could be produced if the productive resources used in the fishery were used for a different purpose. The social opportunity cost is the appropriate cost concept to use when analysing the fishing industry from an aggregate point of view, such as when finding the appropriate rate of exploitation or the appropriate investment in fishing capacity. When markets are perfect, the social opportunity cost will be identical with private cost, as explained in Chapter 1.

It is important to distinguish between long- and short-term costs with respect to social opportunity cost as well as with respect to private cost. Such equipment as nets and fishing boats typically are not useful for purposes other than fishing, and the values forgone by using them for fishing will be small or negligible. This is not so when it comes to the decision to produce nets or fishing boats; the cost of acquiring such equipment is the value of the goods that could be produced with a different kind of equipment that could be acquired instead. Market prices of such equipment and the wages of labour are the most obvious and easily available measures of this cost. The rationale for this is that the equipment or the labour would not be bought or hired unless it could be used to produce a value that covers the cost of doing so. To what extent this is true depends, however, on how well developed the market system is and how accurately the price of equipment and the wages of labour in fact reflect the values that can be produced by these factors.

The *cost of durable equipment* has two components, the cost of the equipment itself and the cost of binding capital in the form of specialized equipment with little or no alternative use. Suppose a piece of equipment is fully financed by a bank loan. The loan itself is equal to what it costs to acquire the equipment. Then there is the interest to be paid on the loan. What does this correspond to? In a developed system of financial markets, the interest that a bank charges on a loan reflects what it could get from loans given for other purposes. The fact that it is at all possible to charge interest on loans is due to the phenomenon that money invested in projects is productive, in the sense that it generates a value that exceeds the cost of the equipment and other inputs necessary to carry the project through. Hence the interest charged on loans reflects a social opportunity cost; that is, the value in excess of costs that could be produced by directing the loan to some other purpose.

The reasoning is no different even if we consider a project that is fully financed by the owner's own funds. If the project is profitable, it must generate a revenue that allows the owner to recover his or her money. Not only that, if the project in question had not been undertaken, the money could have been invested in the capital market where it would have earned interest. The interest forgone would be a part of the cost of the project, just as the interest on the loan would be a cost of a project financed through a bank loan.

Profit is the difference between revenue and cost. Depending on whether we take a long- or a short-term view we can speak about long-term and short-term profit. Looking only at the short term we would subtract only short-term costs from the revenue, so short-term profit would include the contribution to paying for durable equipment like boats. If the short-term profit exceeds this payment, there would be some long-term profit as well. Note that it could be profitable in the short term to operate a fishing boat even if the long-term profit is negative. As long as there is some short-term profit making some contribution to paying the fixed cost that must be paid anyway, it obviously makes sense to operate the boat rather than laying it up. On the other hand it would not be profitable to renew the equipment under such circumstances, as its cost could not be recovered.

Needless to say, second-hand markets for boats must be considered as well. If a boat does not cover its long-term cost while it is possible to sell it for a price that covers more of the long-term cost than is possible by operating the boat, it would obviously be more profitable to sell the boat than to operate it. The problem is, however, that it is not likely that the second-hand market will pay such a high price, unless the new owner believes he will be able to do better than the present owner.

Rent is the amount that could be charged by an owner of a fish stock for using it, just as a landlord charges a rent for letting people live in his house or farm his land. If the owner of a fish stock exploits the stock himself, the rent would be implicit and identical to his profit. If he lets others exploit the stock, the maximum rent he could extract would be equal to the profit earned by those who exploit it, as the sum people would be willing to pay for access to the stock would not be greater than what they expect to have left after all expenses are paid, including the necessary return on their own capital.

We shall in the following make a distinction between profit and rent. If the owner of a resource such as a fish stock controls the market for the product he extracts, he will consider how the market price will be affected when he decides how much to harvest and sell. A part of his profit would thus be due to his market power and not to the ownership of the resource. That part of the profit cannot, therefore, be regarded as rent attributable to the resource but as monopoly rent (excess profit due to market power). Hence the fishing rent is identical with profit only if the owner or the manager of the resource cannot influence the market prices of fish and fishing inputs.

2.4 The free-access fishery

If the exploitation of a fish stock is to be viable in the long term, the amount caught cannot exceed the surplus growth, except temporarily. A useful simplification to discuss this issue is to consider a fishery in a long-term equilibrium; that is, a state in which the amount caught is exactly equal to the surplus growth. This implies that the stock will neither increase nor decrease but remain steady at some value between S^{**} and S^{*}, the minimum viable level and the natural equilibrium level (see again Fig. 2.1). Hence the surplus growth may be said to represent sustainable yield from the stock. The maximum surplus growth and sustainable yield will be attained at some stock level S_{msy}, $S^{**} < S_{msy} < S^{*}$.

Fish stocks typically are (or used to be) free-access resources; that is, anybody has the right to fish from the stock. Before the 200 mile exclusive economic zone became near universal in the late 1970s anyone, irrespective of nationality, had access to most fish stocks of the world, as they were seldom contained within national boundaries. The 200 mile limit has restricted the access to most fish stocks to the citizens of one or a limited number of sovereign states, as relatively few stocks are located outside the economic zone of any state. The citizens of the states in whose zone a stock is located still often have free access to the stock, even if restricted access in one form or another is becoming increasingly common. It is useful, therefore, to begin the economic analysis of fisheries management by considering the free-access fishery. This will establish why there is a need for fishery regulations.

Free-access fisheries are typically characterized by many independent decision units, a decision unit typically being a skipper or a boat owner (who often are the same person), and less frequently a fishing firm with many boats. The essential point of this is that no single decision maker can control the fish stock by his own action. If somebody decides to leave some fish unharvested today in order to take them tomorrow, he has no assurance that they will be there tomorrow as long as others can take them in the meantime. The fisherman who catches fish from a stock to which many others have access thus has a good reason to take as much as he can as quickly as possible. This situation is fundamentally different from agriculture. Apart from the vagaries of the weather, a farmer can be reasonably certain that the corn he leaves unharvested today will still be around and more ripe tomorrow.

It is perhaps obvious that this kind of behaviour leads to undesirable results, but let us look at it in terms of the concepts we have developed. The question we are asking is this: what will be the long-term development of the fishery? To find the answer, we must ask what determines the activity in the fishery in the long term. To make matters as simple as possible, let us assume that fishing requires no special skills and that everybody uses the same fishing technology, so that the fishing fleet consists of identical boats which can be acquired at a constant cost. Since the boats are identical and all fishermen are equally skilled, the catch per boat will be the same for all boats. How many boats will then take part in the fishery? As long as the value of the catch per boat is greater than the cost per boat (both being measured per unit of time), people will find it attractive to invest in new boats. Hence the incentive to invest in new boats will not disappear until the catch per boat has fallen sufficiently to become equal in value to the cost per boat.

By what mechanism will the value of the catch per boat per year be brought into balance with the annual cost per boat? Presumably the fishery developed because the value of the catch per boat was greater than the cost per boat; otherwise the boat owners would not have found it worth while to acquire boats. As long as the value of the catch per boat is greater than the cost per boat, there is an incentive to add more boats. Unless the expansion of the fishery somehow causes the value of the catch per boat to fall, the stock will be depleted beyond recovery.

There are three ways in which the value of the catch per boat could fall as the fishery expands:

(1) the catch per boat might fall as the stock is depleted;
(2) the price of fish might fall as the catch increases;
(3) the cost per boat might increase as the number of boats increases.

Consider (1) first. Suppose that the price of fish and the cost per boat are both constant. Then, if the value of the catch per boat is to fall as the fishery expands and the stock is depleted, the catch per boat must fall. This is the case we dealt with previously when we discussed the concept of fishing effort. If the fish stock is always uniformly distributed over a given area, the catch per unit of effort, or the catch per boat in our present terminology, must fall in proportion to the stock. In reality it may, of course, fall less than proportionately. In the special case when the density of fish is always constant, also discussed above, the catch per boat would not change at all as the stock is depleted.

It is easy to illustrate the long-term equilibrium position of the fishery when the catch per boat varies proportionately with the size of the exploited stock. Denoting the cost per boat by c, the (constant) price of fish by p, and the catch per boat by $f(S)$, a monotonically increasing function as shown in Fig. 2.2, there

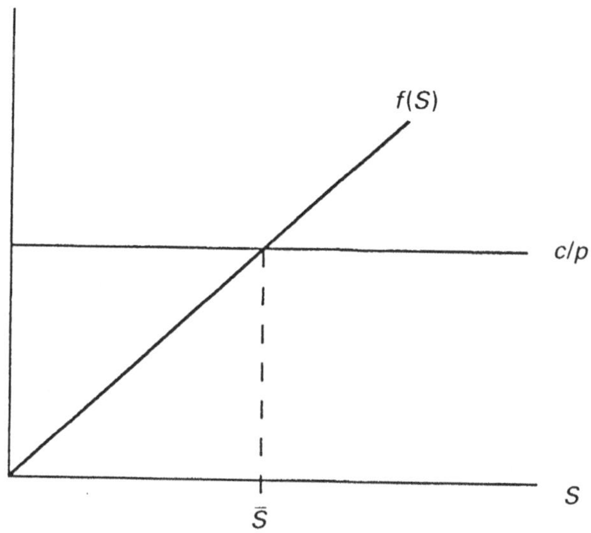

Fig. 2.2 Equilibrium stock ($\bar{S}$): catch per unit of effort [$f(S)$] equals cost per unit of effort (c) divided by the price of fish (p).

will be one particular value of S, $\bar{S}$, compatible with long-term equilibrium, defined as

$$pf(\bar{S}) = c \tag{2.1}$$

The equilibrium is illustrated in Fig. 2.2, and again in Fig. 2.3 as the intersection of the line $\bar{S}$ with the surplus growth curve $G(S)$. The arrow shows the effect of an increased price or a decreased cost per boat on the long-term equilibrium in the fishery. From Equation (2.1) and Fig. 2.2 it is clear that a higher p or a lower c will imply a lower value of $\bar{S}$. This leads to a rather startling conclusion. If the line $\bar{S}$ in Fig. 2.3 is shifted sufficiently far to the left as a result of a higher price or a lower cost per boat, it could quite possibly lie to the left of S^{**}, the smallest viable size of the stock. This means that the stock size that would bring the value of the catch per boat into balance with the cost per boat is too small to be viable. Furthermore, as we have just seen, this is all the more likely to happen the higher the price of fish or the lower the cost per boat. Hence the risk that the fishery will not be sustainable and that it will collapse because of overexploitation will be all that greater the more profitable the fishery is initially. From this we may conclude the common property animal populations will not be exploited in a sustainable way unless it is prohibitively expensive to deplete them below their lowest viable level.

There is thus clearly a need to replace free access with some form of control. Two things which will be elaborated shortly may, however, be noted at this point. First, the need for management does not rest exclusively on the risk of irreversible depletion but can be shown to be necessary in order to attain an economically

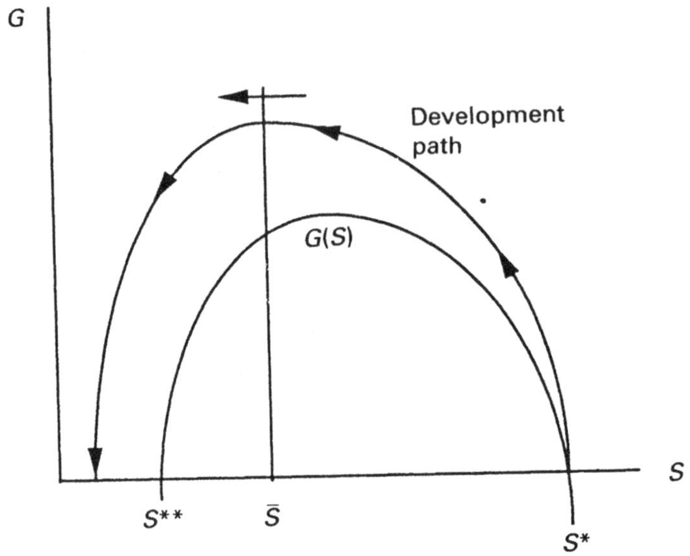

Fig. 2.3 Surplus growth (G), equilibrium stock ($\bar{S}$), and a possible development path that overshoots the equilibrium and leads to extinction.

optimal exploitation. Secondly, it cannot be ruled out that an irreversible depletion could be economically the most desirable strategy of exploitation.

The risk of stock extinction can be much greater than suggested by the above. What if the catch per boat does not respond at all to a falling level of the stock but remains the same as the stock is depleted? With the price of fish and the cost per boat both constant, there will be no mechanism bringing the value of the catch per boat into balance with the cost per boat; investment in new boats will be profitable as long as there are any fish left to be taken. Unfortunately there is some empirical evidence indicating that this is sometimes the case in the real world (Bjørndal 1987). In this case the function $f(S)$ in Fig. 2.2 would be a horizontal line. It would have no point of intersection with the line c/p in Fig. 2.2 (it might coincide with it, however), and the line $\bar{S}$ in Fig. 2.3 would not exist. These points will be further elaborated in Sections 2.7 and 2.8.

Note, finally, that the existence of a long-term equilibrium point given by the intersection between the line $\bar{S}$ and the function $G(S)$ in Fig. 2.3 is not enough to ensure that the fishery will develop in a sustainable way. It is possible that the fishery will react in such a way that the equilibrium will be missed and the fish stock be depleted beyond recovery because of 'over-reaction', like the one shown by the development path in Fig. 2.3. Circumstances that would be likely to lead to such a development are a combination of rapid expansion in response to profits in the fishery and slow exit from the fishery as profits turn to losses. This could be caused by low operating costs of boats that have already been bought, making it profitable to continue to use them for a long time even if their replacement would not be worth while once they are no longer usable. Two examples of fishery development paths, one of which misses the equilibrium, are considered in Section 2.8.

The other two possibilities making a free-access fishery viable ((2) and (3)) may be briefly mentioned. If the price of fish falls as the catch increases, it may bring about a balance between the value of the catch per boat and the cost of a boat. Whether or not this will happen depends on how the catch per boat changes as the fishery develops and how sensitive the price is to changes in the catch volume. Suppose the stock is greater than S_{msy} (see Fig. 2.1), so that the sustainable yield increases as the stock is reduced. As the catch increases the price falls, and the value of the catch per boat would fall even if the catch per boat is constant. This mechanism may, however, work the other way. Suppose that the value of the catch per boat is still greater than the cost per boat when the stock has been depleted to S_{msy}. An expansion of fishing effort would reduce the stock further, the catch would fall in the long term, and the price would rise. If the price is sufficiently sensitive to changes in the catch volume, the catch per boat would rise in value terms even if it falls in volume terms. Unless the unit cost of additional boats increases even more, this would invite a further expansion of effort until the stock had been totally depleted.

The case of an increasing cost per boat takes us beyond the assumption we made for convenience that all boats are identical and can be acquired at a constant cost. In reality this need not be true. By a more general approach it can be shown that an increasing cost per boat, or rather fishing effort, increases the likelihood that a long-term equilibrium with a viable stock will exist. This and other technical issues related to the existence and attainment of free access equilibrium are discussed in greater detail in Section 2.8. The implications of an increasing cost per boat for free access equilibrium will be further considered in Section 2.7.1.

2.5 The social objective of fishing

What is the purpose of fishing? For the individual fisherman this is easy to answer. If asked, he would probably say 'to make as decent living as possible' or something like that. From the point of view of the economy as a whole the answer is similar, even if the perspective is slightly different. The fishery is only one of many activities that contribute to the *well-being* of a nation. At each point in time the nation has a given amount of resources at its disposal; manpower, production capital such as buildings, equipment, etc., and natural resources. These resources can be put to various uses, and the task is to do so as efficiently as possible, so as to provide the highest possible level of well-being, including that which must be set aside for economic growth (enhancement of well-being in the future).

The term well-being is a fairly imprecise concept, and it certainly has both material and immaterial aspects. In order to make it operational it must be made more precise. The fishery (excluding fishing as a leisure activity) contributes primarily to material well-being through providing food supplies or foreign currency, which in turn may be used to buy other commodities that contribute to material well-being.

The question is, then, how much of the productive disposable resources should be allocated to the fishery rather than to other industries that also contribute to the material well-being through food supplies or in other ways. First of all, we need to measure the contributions by the various sectors of the economy by some common denominator. If the task is as simple as providing nutrition, this could possibly be done by measuring the provision of proteins, but even that is doubtful since we do not live by proteins alone. The common denominator usually employed for this purpose is money, on the assumption that the price of a commodity provides information about how much a unit of the commodity is worth, in terms of material well-being, in relation to other commodities. If a kilogram of meat costs twice as much as a kilogram of fish, this means that an additional kilogram of meat contributes twice as much to the material well-being of consumers as an additional kilogram of fish, for otherwise they would not buy these commodities at these prices (see again the discussion in Chapter 1). The reason need not be that meat is twice as nutritious; it could very well be because meat is a scarcer commodity and therefore especially cherished. The relative prices of commodities typically are sensitive to the relative scarcities of commodities; in Saudi Arabia a litre of water costs more than a litre of petrol, but not so in other oil producing countries with different climates, like Britain and Norway.

As discussed in Chapter 1, the cost to the economy of providing an additional amount of fish is the value of other commodities that could be produced instead. It would, however, be a cumbersome procedure to try to calculate the value of the commodities that could be produced instead of fish. Instead one uses the cost of the inputs that are used in the fishery, labour costs, the cost of boats and equipment, fuel, etc. This is, however, correct only if these costs measure correctly the value that could be produced with these inputs if they were used in a different way. As discussed in Chapter 1, there is a presumption that this will be true in a market economy.

The market prices of labour, fuel and materials are therefore useful reference values in accounting for the costs of fishing. One should, however, look for market imperfections that might cause deviations between these prices and the values that these inputs could produce in other sectors of the economy. The wages of labour are particularly likely to have to be corrected for such differences. There are two reasons for this. Firstly, there is the fact that fishermen typically are hired on a share basis, that is, not for a given wage but for a share in the catch value. As fishermen thus carry some of the risk of fishing, their remuneration is likely to be higher than what they could get elsewhere and the value they would produce in other sectors of the economy.

Secondly, successful fishing requires considerable technical skills. These skills are, at least to some extent, acquired and not necessarily useful in other sectors where comparable skills are similarly acquired through training and experience. The short-term gains of transferring labour from the fishing industry to other occupations are likely, therefore, to be much lower than the incomes normally obtained in these occupations. In the long term this will not be so, as fishermen will gradually acquire new skills and those who might otherwise have become fishermen are employed by other industries. When accounting the costs of the fishery one will therefore have to make a distinction between long-term and short-term decisions; when deciding the long-term capacity of a fishery it would be correct to disregard the special skills needed in fishing and to account the costs of labour as if its skills were freely transferable from one sector to another, while when taking short-term decisions (how much to fish and how many boats to use any particular year) the non-transferability of skills would have to be taken into account.

A similar consideration applies to the use of specialized equipment. Such equipment (boats, engines, electronic devices) typically has very little use in other sectors of the economy, and hence the short-term cost of such equipment is low or negligible. The situation is different with respect to whether or not to acquire such equipment. In that case it would be reasonable to use market prices for accounting the costs of such equipment, unless there are valid reasons to believe that these prices are distorted in such a way that they do not reflect the value that the same amount of money could produce if used in other sectors of the economy. Examples of this are soft loans and other subsidies for building fishing boats. Such arrangements imply that the buyer of the boat is not paying its full cost; some is being paid from public funds. If the money (the cost of the boat including the subsidies) were used instead to buy some other non-subsidized equipment used in other industries, it would presumably produce a value at least equal to the cost of that equipment. Hence the amount that the fisherman pays for the boat does not fully reflect the social opportunity cost of the boat.

Having found a common measuring rod for the costs and benefits of fishing, we can answer the question how much of the productive resources at disposal should be used in the fishery. Clearly, as long as an incremental unit of a factor of production produces a greater value in the fishery than elsewhere in the economy, the amount allocated to the fishery should be increased. For various reasons, not least the limited productivity of fish stocks, the contribution of an additional unit of a factor of production allocated to the fishery is likely to diminish as more is being used in the fishery. Hence the use of factors of production in the fishery should proceed to the point where the value produced by the last unit is equal to the value it could produce in other industries.

The use of various factors of production in the fishery raises questions as to how they should be combined; that is, what kind of technology to use. We shall not discuss these issues here in any detail, but proceed on the basis that the most cost-efficient technology has been chosen. By cost-efficient technology we mean the combination of factors of production that produces a given amount of fish at the lowest possible cost. An offshore trawler versus an inshore boat would be examples of two technologies, with the trawler using relatively little manpower and requiring a large initial investment, compared with the inshore boat. In economies where manpower is scarce and wages accordingly high, the trawler would probably represent the most cost-efficient technology, as it uses relatively little of the scarce factor of production.

Assuming that a single technology is the most cost-efficient implies that the fishing fleet should consist of identical vessels. Even if each vessel represents a combination of various factors of production, the fact that the combination is the same for all vessels implies that they can be represented by a single factor of production, or index, which we shall call fishing effort. In line with our previous discussion of this concept, fishing effort means the application for a certain period of time of a certain number of vessels.

The optimal effort in the fishery can be found by maximizing the resource rent in the fishery. This may appear strange. Suppose a fish stock is owned by some individual or company which leases the right to use it to others. The rent the stock can provide would accrue to the owner. It is not at all obvious why a maximization of this income should be a social objective, particularly since it is earned by ownership alone and not by any productive effort. The maximization of rent is not, however, an objective in itself, but a means to achieve the best possible balance between the use of resources in the fishery and the rest of the economy. At the point of optimum effort, consumers' willingness to pay for an additional unit of fish will be equal to the cost of providing that unit. This is precisely the condition for maximum rent, as is shown mathematically in Section 2.8.

It is, however, possible to make the maximization of rent more meaningful as a social objective. Suppose the fish stock is a public resource, managed by the public authorities for the benefit of the public. In that case the fishing rent would accrue to the public purse and could be used for the public benefit. Even if the fish stock were privately owned and the rent were collected on the first round by private individuals or firms, it could still be in part used in this way, as it could be subject to a special tax. This is typically done with respect to natural resources such as minerals and petroleum.

Figure 2.4 illustrates the implications of maximizing fishing rent and compares it with maximizing profits in the case when the owner of a fish stock controls the price of fish. The curve $p(Y)$ shows how the price (p) varies with the quantity caught (Y), while the curve MC shows the marginal cost of providing the fish, including the impact of an increased catch on future catches and costs. Maximizing the difference between the aggregate revenue and the aggregate cost implies a catch volume where the marginal revenue (the revenue of an incremental catch of fish) is equal to the marginal cost of catching the fish. How, then, does the

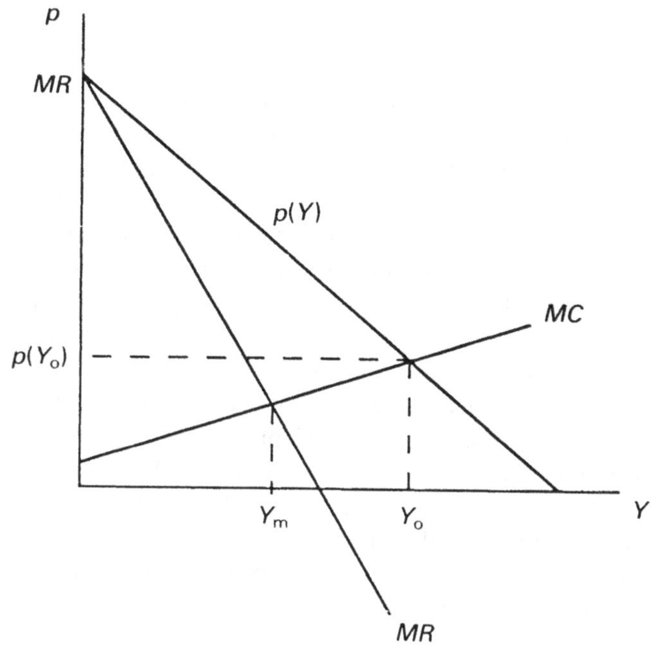

Fig. 2.4 Socially optimal catch (Y_o) versus optimal catch for a monopolist (Y_m).

marginal revenue vary with the catch volume? The first unit of fish that is sold will increase the revenue from zero to a value equal to the price of that first unit, so the marginal revenue of the first unit will be equal to the price of that unit. Now, in order to sell one more unit the price will have to be lowered, unless it is possible to sell two units of the same commodity at different prices. The marginal revenue of the second unit will therefore be equal to the price of that unit less the necessary reduction in the price of the first unit. From this we see that the marginal revenue of all other units than the first will be less than the price at which the marginal unit can be sold. For the demand function in Fig. 2.4 the marginal revenue curve will be a straight line but twice as steep as the demand curve, as shown by the line *MR*. This can be seen as follows. The revenue is $R = aY - bY^2$. The marginal revenue is the derivative of R, which gives $MR = a - 2bY$.

The catch volume that would maximize the aggregate profit is Y_m, where the marginal revenue is equal to the marginal cost. At this point the price is higher than the marginal cost. Since the price expresses consumers' willingness to pay for an incremental unit, this would imply that consumers are willing to pay more for an additional unit of fish than it costs to increase the supply. Hence the quantity Y_m cannot be the optimal one, from a social point of view. The optimal quantity is Y_o, where the demand curve intersects the marginal cost curve. At this point consumers' willingness to pay for an incremental unit is equal to the cost of that unit.

We would arrive at the point Y_o if we maximized the sum of differences between consumers' willingness to pay for a unit of a commodity and the cost of providing each unit. In Fig. 2.4 this corresponds to the area between the demand curve and the marginal cost curve. The commonsense interpretation of this is that we would be maximizing the net benefit of providing the commodity. There is a name for it; it is called consumer and producer surplus. Consumer surplus is the difference between what the consumer is willing to pay for a commodity and what he in fact pays, the latter being the price for the marginal unit, provided all units are sold at a uniform price. Producer surplus is the difference between what a producer gets for a commodity, which is equal to the price of the marginal unit, provided that all units are sold at a uniform price, and the cost of production. In Fig. 2.4 the consumer surplus of the optimal quantity Y_o is the area between the demand curve and the price line $p(Y_o)$, while the producer surplus is the area between the said price line and the marginal cost curve. At Y_o the sum of these is maximized. In Section 2.8 it is shown that maximization of the sum of consumer and producer surplus gives the same result as maximization of the fishing rent (the difference between revenue and cost, with the catch being valued at a price equal to the price of the marginal unit at the optimum).

Does it never make sense to maximize the aggregate profit, taking into account how the price depends on the quantity sold? For a sole producer who totally dominates the market for fish it would, but from the point of view of society as a whole it would not, unless we are dealing with a country that exports its fish to other countries in exchange for other goods. In such a situation the country is in a position similar to the monopolist who maximizes his benefit by manipulating the market price to his advantage.

2.6 The sustainable yield curve

In Fig. 2.5 a relationship is derived between sustained fishing effort and sustainable yield in the case when catch per unit of effort is proportional to the size of the fish stock. In the left half of the figure the surplus growth curve, $G(S)$, is shown, together with lines showing how the catch varies with the stock for a given level of effort (Z). The straight shape of these lines implies that a given level of effort will catch a constant proportion of the stock, as the catch increases proportionately to the stock for a given level of effort. Increasing the effort from Z_1 to Z_2, say, would increase the catch for any given level of the stock (the constant q implies that the catch increases proportionately to effort for any given level of the stock). As explained in Section 2.2, this relationship between stock and catch would hold, for any given level of effort, when the stock is always uniformly distributed over a given area.

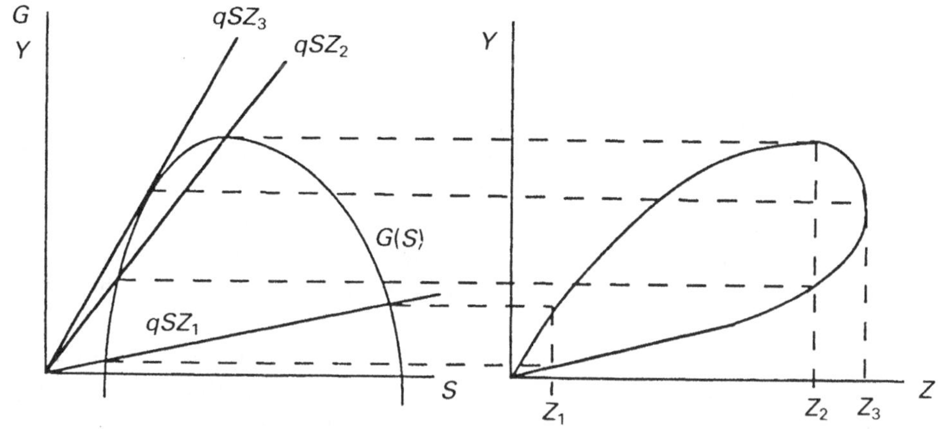

Fig. 2.5 Sustainable yield (Y) as a function of effort (Z) derived from the equality between surplus growth (G) and catch (qSZ).

To derive the relationship between sustained effort and sustainable yield, we note that the yield must on average be equal to the surplus growth in order to be sustainable. We shall simplify this to the point of requiring that the catch must be kept constant at a level equal to the surplus growth. In this simple model where the rate of growth depends only on the size of the stock, this implies that the stock is being kept at a certain level yielding a rate of surplus growth equal to the amount harvested per unit of time by a steady level of effort. Hence all possible sustainable yields are given by the points of intersection between the catch lines and the surplus growth curve in the left half of Fig. 2.5. By looking at these points of intersection we see that the sustainable yield increases with fishing effort up to a certain point and then decreases again. Furthermore we see that each level of effort is associated with two different levels of sustainable yield. The sustainable yield curve (the relationship between sustained effort and sustainable yield) will therefore look like the curve in the right half of Fig. 2.5; it will rise to a maximum, fall off to the right and then bend under itself back to the origin.

The case in which the catch increases proportionately to the stock is somewhat special. Even if it is quite likely that the catch will increase with the stock for a given level of effort, the relationship need not be proportional. It is quite likely, for example, that the catch will increase with the stock at a decreasing rate. In that case the lines shown in the left half of Fig. 2.5 would not be straight but bent and approach a horizontal asymptote. This would not affect the shape of the sustainable yield curve in any fundamental way. If, on the other hand, the growth curve starts at the origin and each catch line (or curve) has only one point of intersection with it, the backward-bending part of the yield curve would be eliminated, and it would have a regular shape like the curve in Fig. 2.7.

The backward bending part of the sustainable yield curve is of limited significance, as all the points on it are unstable. To see why, consider a given catch line in the left half of Fig. 2.5. At any point outside the growth curve the catch exceeds the surplus growth, the stock diminishes, and we slide down the line, as it were. At any point inside the curve the catch is less than the surplus growth, the stock increases, and we move up the line. From this we see that if a catch line has two points of intersection with the growth curve, the left one must be unstable, as the slightest deviation would move us further away from it, either upwards towards the stable point, or downwards towards extinction of the stock. The left points of intersection correspond to the backward bending portion of the sustainable yield curve, as is clear from Fig. 2.5.

The case in which the catch depends only on effort and is totally independent of the stock (except that there must be some fish available if anything is to be caught at all) is sufficiently different to merit a special discussion. Recall that this is the case in which the density of fish is always constant, so that the size of the area a stock occupies is directly proportional to its size. The derivation of the sustainable yield curve is shown in Fig. 2.6. As in Fig. 2.5, the left half shows the surplus growth curve and the catch for given levels of effort. Since the catch is independent of the stock level, the lines showing the catch taken by given levels of effort are horizontal. The resulting sustainable yield curve is shown in the right part of Fig. 2.6. It will be a straight line if the catch per unit of effort is always the same, whereas it will rise at a decreasing rate up to its maximum if the catch per unit of effort decreases as more effort is applied, like the dotted line shown in the right part of the figure.

2.7 Optimum exploitation of fish stocks: static, dynamic and stochastic models

We shall discuss the optimal utilization of fish stocks in the following three contexts:

(1) deterministic models without discounting (static models);
(2) dynamic deterministic models;
(3) stochastic models.

In deterministic models without discounting we are concerned only with long-term effects. As has been explained above, fishing involves a deviation from an original, pristine situation; in natural equilibrium there is no surplus growth, whereas a reduction of the stock generates some surplus growth that can be harvested. During the phase of transition from a pristine equilibrium without fishing to a new equilibrium with a sustainable level of fishing, it is possible that the catch and the rent will be greater than can be sustained in the long term. If future rent is not discounted in relation to current rent, then clearly all that matters is the long-term sustainable rent; no trade-off between a temporary gain

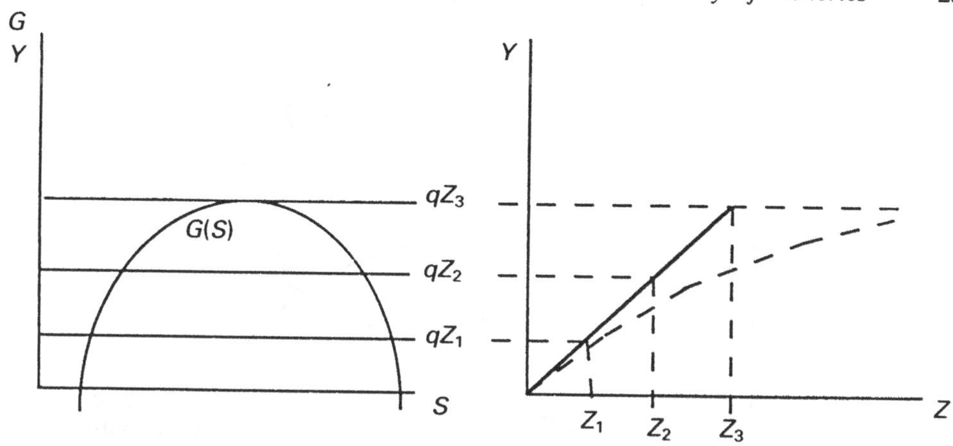

Fig. 2.6 Sustainable yield (Y) as a function of effort (Z) when the catch per unit of effort is independent of the stock. The dashed line shows sustainable yield when the catch per unit of effort decreases as the effort increases.

and a permanent loss could ever make sense, since the latter would always dwarf the former simply by extending an annual loss, however small, into the indefinite future. This is case (1) above.

Dynamic deterministic models deal with the trade-off between present and future rent. If future rent is discounted in relation to current rent, it could make sense to incur some loss of future sustained rent for a gain in current rent. Even if a sustained loss must always be greater than a temporary gain, this is no longer so when we discount systematically long-term effects, the heavier the further into the future they occur. This is case (2).

As already discussed, the growth of fish stocks is determined not only by the rate of exploitation through its effect on the size of the stock: environmental effects (ocean currents and temperature, rainfall, size of prey and predator stocks) also play an important role. For this reason, 'sustainable' yield of fish stocks will not really be constant over time; it is likely that the catch will have to be varied in response to the fluctuations in natural conditions. A sustainable yield is therefore to be understood as a long-term average around which the actual catch will fluctuate. This is case (3) above.

In this chapter we discuss in particular cases (1) and (2), and mathematical details are shown in Section 2.8. We shall offer some preliminary considerations on case (3), but this will be treated more fully in Chapter 4.

2.7.1 *The static case*

As explained in Section 2.5, the optimum rate of exploitation implies maximization of rent, which is equal to profit when the catch is valued at a price equal to

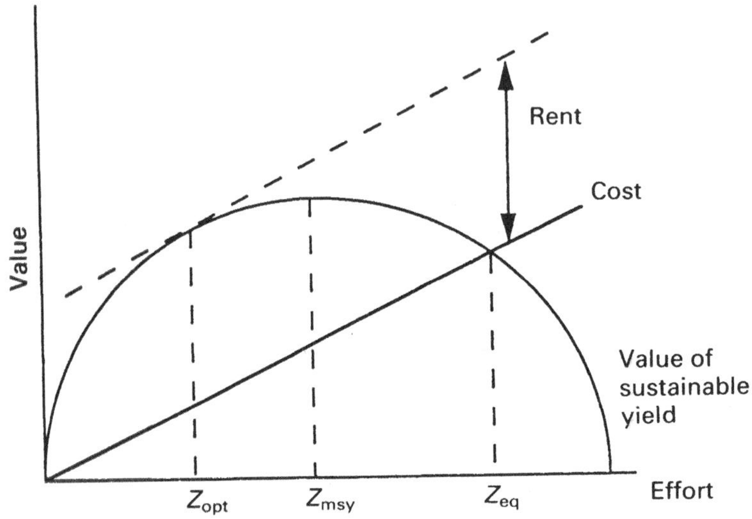

Fig. 2.7 Free access equilibrium effort (Z_{eq}) and optimum effort (Z_{opt}) when the future is not discounted. Catch per unit of effort proportionate to the size of the stock.

the willingness to pay for the marginal unit at the optimum. A greater fishing effort means more and possibly more intensive use of manpower, boats and other inputs, and so the total cost of fishing must increase with effort. To begin with we shall assume that the cost increases proportionately to effort. This gives rise to the cost lines shown in Figs 2.7 and 2.8.

Multiplying each point on the sustainable yield curve by a constant price produces a sustainable revenue curve having the same general shape as the sustainable yield curve. The curves obtained in the two cases with a constant versus stock-dependent catch per unit of effort are shown in Figs 2.7 and 2.8. In the case of stock-dependent catch per unit of effort (Fig. 2.7) we have assumed that the surplus growth curve starts at the origin and that each 'catch line' (see Fig. 2.5) intersects it only once.

The sustainable rent is maximized where the distance between the cost line and the sustainable revenue curve is the greatest possible. When the catch per unit of effort depends on the stock (Fig. 2.7), the optimum level of effort (Z_{opt}) is less than needed to take the maximum sustainable yield (Z_{msy}), and even further below the equilibrium free access effort (Z_{eq}), where total cost equals total revenue, as explained in Section 2.4. Hence we may conclude that fish stocks will not be optimally exploited in free-access equilibrium. It furthermore appears that economic optimization is more conservation minded than fishing based on maximizing the sustainable yield, as the economically optimal effort is less than the effort needed to maximize the sustainable yield. This is not, however, a generally valid conclusion, as will be demonstrated in connection with dynamic optimization below.

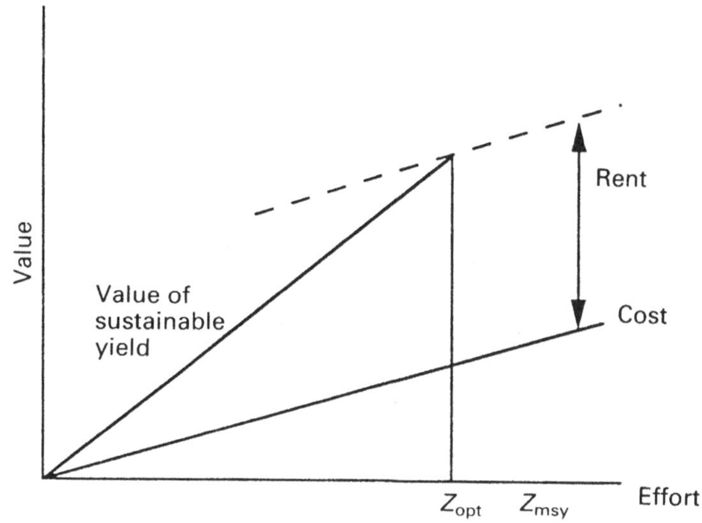

Fig. 2.8 Optimum effort (Z_{opt}) when the future is not discounted. Constant catch per unit of effort.

Figure 2.8 shows cost and sustainable revenue when the catch per unit of effort is constant and independent of the stock. Here the maximum sustainable yield coincides with the economically optimal yield. From Section 2.4 it may be recalled that in this case there will be no equilibrium under free access. From Fig. 2.8 we see that there is no point of intersection between the revenue curve and the cost curve; the fishery will always yield some excess profit, and nothing will stop a competitive free-access fishery from expanding until the fish stock has been wiped out. Clearly this is not an economically optimal policy, on the basis of the assumptions made thus far.

The conclusion that the maximum sustainable yield is also the economically optimal yield will have to be modified if the price of fish is so sensitive to changes in the quantity caught that it would be lower than the unit cost of fish at the maximum sustainable yield level. Consumers' willingness to pay for the last unit of fish would then be insufficient to cover the cost of that unit, and there would be no point in pushing the yield that far. The optimum yield would then be the quantity where the price (the willingness to pay for the marginal unit) would be equal to the unit cost. In this case the free-access equilibrium would in fact exist and coincide with the economic optimum. The rent in the fishery would be zero, reflecting the fact that the stock would have no scarcity value. There would be more than enough surplus growth to satisfy the demand that it is worth while satisfying, and the stock would have no value as a factor of production because the catch per unit of effort is constant and independent of the stock.

What about progressively rising total cost of effort? Consider, first, the case in which the catch per unit of effort is proportionate to the stock. This is shown in Fig. 2.9. The conclusions arrived at are only slightly different from those obtained

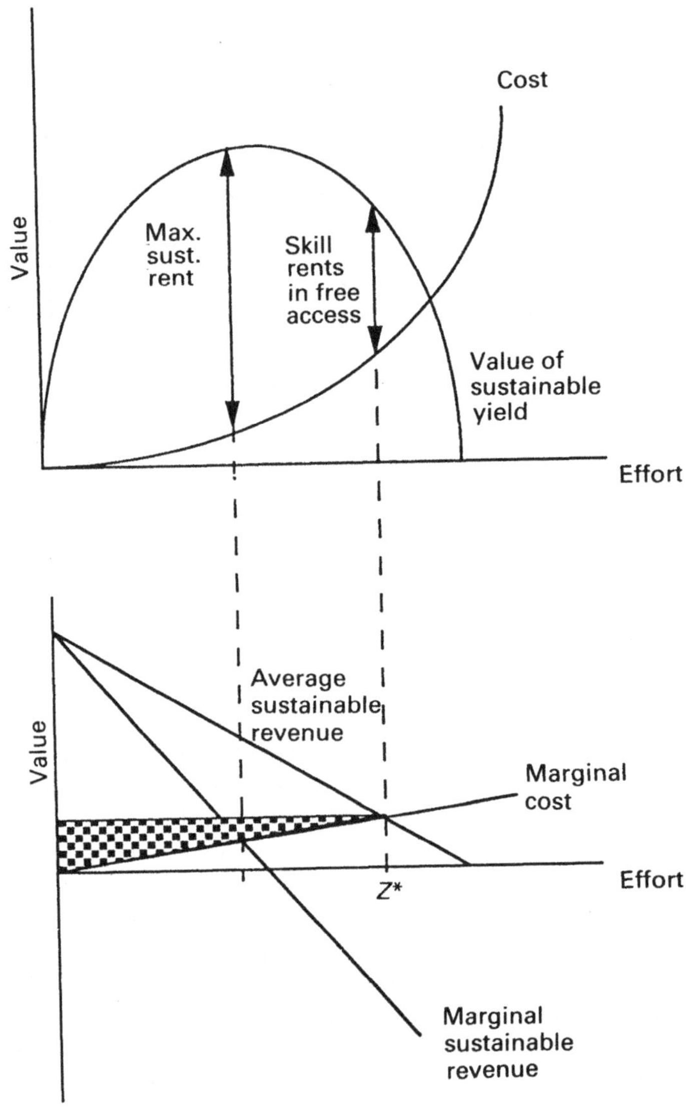

Fig. 2.9 Comparison of optimum effort and free-access equilibrium effort (Z^*) when the marginal cost of effort rises with effort. Some skill or outfit rents are earned in free-access equilibrium (shaded area). Catch per unit of effort increases with the stock. No discounting of the future.

earlier. The optimum effort is still where the distance between the revenue curve and the cost curve is the greatest possible, and it is seen to be less than the maximum sustainable yield effort. It is with respect to the absence of any long-term profit under free access that the conclusions differ from the earlier ones. The lower part of Fig. 2.9 shows (1) the sustainable revenue per unit of effort, (2) the marginal sustainable revenue of effort (that is, the incremental revenue resulting from a small increase in effort), and (3) the marginal cost of effort. Because the total cost of effort rises progressively, the cost per unit of effort must be rising, and so must the marginal cost of effort (the cost of an additional unit of effort). The rent will be maximized where the marginal cost of effort equals the marginal revenue of effort. In free-access equilibrium the marginal cost of effort would be equal to the average revenue of effort, because all units of effort yield the same revenue, but the last unit must break even in order that there be no incentive to increase effort. The total cost of any given level of effort such as Z^* is the sum of the marginal costs up to that level, while the total revenue is the average revenue at that level times the level of effort. We see that there must be a difference between the total revenue and the total cost of effort Z^* equal to the shaded triangle in the lower part of Fig. 2.9. This implies that all units of effort except the marginal one (the unit at Z^*) will obtain some profit, because they can be provided at a lower cost than the marginal unit while all units earn the same revenue. Since the marginal unit at Z^* earns no profit, Z^* is the equilibrium effort level under free access.

The reason why some units of effort are cheaper than others could be differences in skill, technology, or equipment, as effort must be measured in homogeneous, or equivalent, units. The profit arising from this is called intra-marginal rent, skill rent, or outfit rent. To the extent that such rents are due to skills or outfit that is easily acquired, they cannot be expected to remain positive for long.

For the case of a constant catch per unit of effort we refer back to Fig. 2.8. With a rising marginal cost of effort, the cost curve will not be a straight line but bend upwards, like the cost curve in Fig. 2.9. This makes it more likely that the marginal cost of effort will become equal to the value of the catch per unit of effort at a catch and effort level which does not exceed the maximum that the stock can support. If such an equilibrium exists, some aggregate profit will be attained in equilibrium, just as in the previous case.

2.7.2 *The dynamic case*

Earlier we described how the exploitation of previously unutilized, or underutilized, stocks may be divided into two phases, the first phase involving a depletion of the stock to a level giving the optimal sustainable yield, at which it is kept for ever after. When a dollar (or whatever) of long-term benefit counts for the same as a dollar of short-term benefit, it is sufficient to consider only the rent accruing once the optimum stock level has been attained. The optimum stock level will by assumption be maintained into the indefinite future, and hence it cannot possibly be desirable to incur any loss of sustained rent in order to obtain a short-term gain.

The situation is quite different if long-term benefits are systematically undervalued, or discounted, in relation to short-term benefits. Let us consider again the two phases of exploiting a previously unexploited stock. Suppose we have attained a stock level which gives the maximum sustainable rent. Would it be desirable to deplete the stock even further? If we did so, our short-term gain would be the rent of the additional catch we could take. Our long-term loss would be the loss of sustained rent implied by moving to a non-rent-maximizing sustainable stock and yield. If these two effects cancel, we have found the optimal stock level.

How does this 'intertemporally optimal' stock level differ from the one previously found? Let us look again at the two cases discussed above, (1) that in which the catch per unit of effort is proportional to the size of the exploited stock, and (2) the other, where it is constant. Consider, first, the immediate gain. Suppose the optimal size of the stock has been attained and consider a small deviation from this optimal level. Denote this change in the stock by ΔS. The short-term gain from this will be $-(p - c_y)\Delta S$, where p is the price of fish, c_y is the cost per unit of fish, and $-\Delta S$ is the temporary change in the catch needed to change the stock by ΔS.

The long-term loss per unit of time will be equal to the change in sustainable yield resulting from the said adjustment of the stock level. The rate of growth, or growth per unit of time, depends according to our previous assumptions on the size of the stock, a relationship which we have denoted by $G(S)$. Now let the change in the rate of growth resulting from a small adjustment of the stock be denoted by $G'(S)$, measured in relation to the change in the stock. The meaning of this is as follows. If G is, say, 100 000 t per year, and an increase in the standing stock by 10 000 t would increase the growth to 105 000 t per year, then G' would be equal to 0.5.

The value of the change in growth and sustainable yield resulting from a stock adjustment of ΔS would thus be equal to $(p - c_y)G'(S)\Delta S$ per period of time. To get the present value of the total long-term change we must calculate the sum

$$X = ax + a^2x + \ldots + a^tx + \ldots$$

where $x = (p - c_y)G'(S)\Delta S$, and $a = 1/(1 + r)$, where r is the rate of discount. Using the formula for convergent geometric series we get

$$X = x/r$$

where X is the so-called present value of the infinite stream of xs. Note that the present value is finite even if the infinite stream of xs is infinite, owing to the discounting of future rents. The basis of this procedure is that one unit of money today is worth more than the same unit a year from now, as long as a positive rate of interest can be earned on investing money. Suppose person A makes a promise to person B to pay 100 units of currency a year from now. How much would a third person, C, be willing to pay B for this promise, assuming that there is no doubt whatsoever that it will be kept? If C does not buy the promise, he could

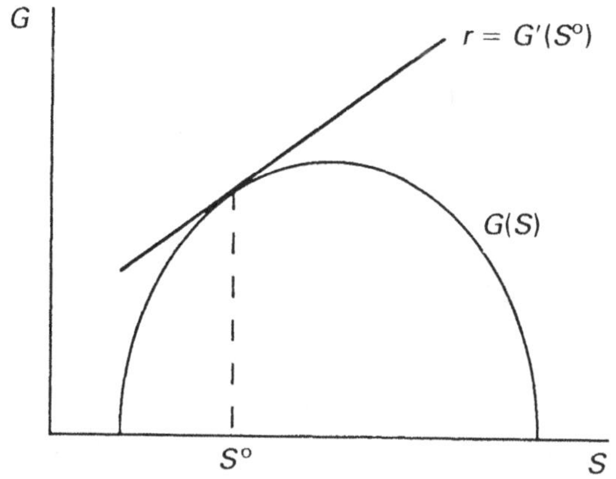

Fig. 2.10 Optimum stock level (S^o) when the future is discounted. Constant catch per unit of effort.

put his money in a bank and earn an interest of r times 100 per cent over the year. In one year's time a sum X would have grown to $X(1 + r)$. If C buys the promise, he will get 100 a year from now, so he will not be prepared to pay more than the sum which, with interest, would pay him 100 a year from now. This amount is equal to $100/(1 + r)$, which is called the present value of 100 units of money a year from now.

Even if the harvesting of fish may be said to be more serious than such money games, the same principle applies. The rate of interest that can be earned in the banks may be expected to reflect the rate of return on investment in the economy in general. Harvesting more fish today and investing the income gained by this in productive processes of some kind is exactly analogous to putting money in the bank, and in market economies this is exactly how many individuals and firms put their savings to work and generate growth in the economy.

We are now ready to state the condition for the optimal stock level in the case when the catch per unit of effort is constant and independent of the stock. The condition is that short-term gains (losses) and long-term losses (gains) from a marginal adjustment of the stock should cancel. Adding the short- and long-term effects derived above and setting the result equal to zero gives

$$-(p - c_y)\Delta S + (p - c_y)G'(S^o)\Delta S/r = 0$$

where S^o denotes the optimal stock level. This may also be written as

$$G'(S^o) = r \qquad (2.2)$$

Figure 2.10 illustrates the optimal solution. The term $G'(S^o)$ is equal to the slope of the tangent to the curve $G(S)$ at S^o. We see that the optimal stock level is less than that which would provide maximum sustainable yield and which we previously found (Section 2.7.1) would be economically optimal in this particular case (constant price and constant catch per unit of effort). Hence some biological

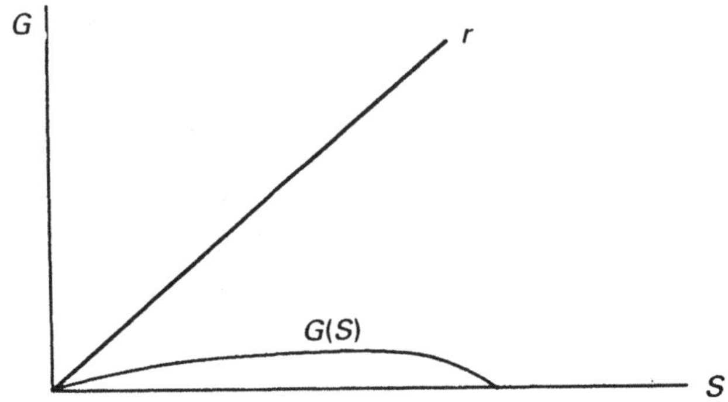

Fig. 2.11 A case where extinction is optimal because the marginal rate of growth $[G'(S)]$ is always less than the rate of interest (r).

overexploitation (that is, driving the stock below the maximum sustainable yield level) could be justified, because the short-term gain would outweigh the present value of future sacrifices.

In fact much more dramatic biological overexploitation could be justified on economic grounds. In order to get an optimal stock level that is viable, Equation (2.2) must have a solution at a viable stock level. Figure 2.11 illustrates a case where this does not happen. The highest possible value of $G'(S)$ is less than the rate of discount (r). In this case it would be most profitable to fish the stock to extinction. Needless to say, such drastic changes call for a more careful evaluation than is obtained by looking only at catches of fish. Fish stocks may have an economic value as such, and the option of being able to fish from stocks that are currently of little interest is of some value.

To analyse similarly the case in which the catch per unit of effort is proportional to the stock, we need to take into account that the cost per unit caught will in this case change with the stock. Instead of a constant cost per unit caught we have a function $c_y(S)$. In line with the concepts developed above, let the change in the cost per unit of fish caught resulting from a 'small' change in the stock be $c_y'(S)$ per unit of change in the stock. The total change in long-term (sustained) fishing costs resulting from a change of ΔS in the stock will then be given by multiplying $c_y'(S)$ by the sustainable yield, which is equal to the surplus growth, and the change in the stock:

$$c_y'(S^o)G(S^o)\Delta S$$

This must be subtracted from the change in the sustainable catch value, for if this term is negative it means that costs are being saved, while in the opposite case costs are being increased. Instead of Equation (2.2) we get

$$-[p - c_y(S^o)]\Delta S + \{[p - c_y(S^o)]G'(S^o) - c_y'(S^o)G(S^o)\}\Delta S/r = 0$$

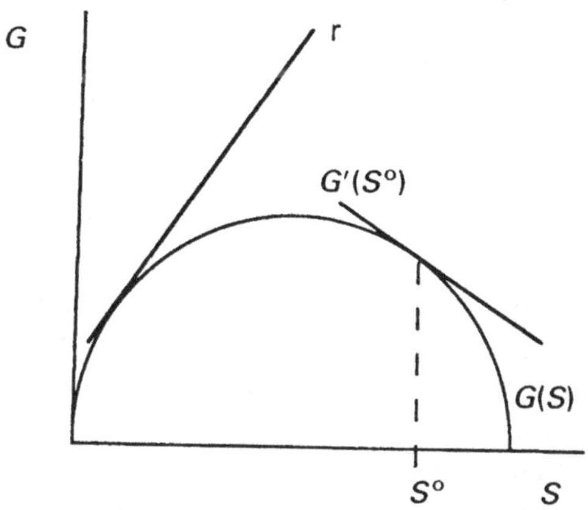

Fig. 2.12 Optimum stock level (S^o) when the catch per unit of effort increases with the stock and the discount rate (r) is low.

which may be written as

$$\{r - G'(S^o)\}[p - c_y(S^o)] = -c_y'(S^o)G(S^o) \tag{2.3}$$

Since the catch per unit of effort increases with the stock, the unit cost must decrease as the stock increases ($c_y'(S) < 0$), so the right-hand side is positive. This implies that $r > G'(S^o)$. Hence, if r is sufficiently small, $G'(S^o)$ must be negative. In that case the optimal stock is greater than needed to support the maximum sustainable yield and the sustained effort is less than needed to take the sustainable yield, as in the static case in Section 2.7.1. This solution is illustrated in Fig. 2.12.

Two special cases may be mentioned. If the unit cost is insensitive to the stock level, the term on the right-hand side will be zero, and we are back with the case considered above where $r = G'(S^o)$. If the discount rate is zero, we get

$$G'(S^o)[p - c_y] = c_y'(S^o)G(S^o)$$

and, since the right-hand side is negative and $p - c_y > 0$, the optimum stock must be greater than needed to support the maximum sustainable yield [$G'(S^o) < 0$]. Here we are in fact back with the case in which only the sustainable rent counts. The conditions for optimal use of fish stocks are derived formally in Section 2.8.

2.7.3 *The stochastic case*

As discussed earlier, the growth of a fish stock depends not only on its size but on other factors as well. The survival rates of eggs and larvae are known to vary a great deal, depending on environmental conditions such as sea temperature, currents, etc. In age-structured populations this gives rise to variations in recruitment of new year classes and hence to variations in the exploitable biomass. The growth rate of individual fish also appears to vary considerably in response to environmental conditions, with similar consequences for the exploitable biomass. Changes in environmental conditions may cause complicated fluctuations in the marine ecosystem, as they are transmitted via changes in the abundance of prey species to changes in the abundance of species that feed on them.

These fluctuations imply that the decision on how much fish to catch in any particular period is not as simple as finding once and for all the optimal level of the stock and then harvesting the surplus growth. The surplus growth of a stock of a given size will vary from year to year, according to the environmental conditions prevailing at each particular time.

To make the right decision about how much fish to catch and how much effort to apply in any given period, it is necessary to estimate how great the surplus growth will be in any given period and to evaluate the consequences of varying the catch in the current period on the catch possibilities in future periods. Since the causes of the variations in fish growth and recruitment are poorly understood and may depend on unpredictable factors, this involves planning under uncertainty. One aspect of this relates to finding the optimum level of investment in fishing capacity. Fishing requires investment in equipment that is not always useful for purposes other than fishing; sometimes it is useful only in a very special type of fishing. Much of the cost of fishing therefore is fixed cost which must be paid whether the equipment can be used or not. This is important, because the ability to utilize fully exceptionally good catch possibilities may require investment in equipment for which there will be no use even in normal years. Such extra investment must be carefully planned in order to avoid incurring irreversible costs that cannot be recovered.

The extent to which this problem is relevant depends both on the variability of allowable catches and on the interdependence between the catch per unit of effort and the size of the stock. If the catch per unit of effort is proportional to the size of the stock, the fishing capacity of a given fleet (level of effort) becomes elastic; that is, it will vary proportionately to the size of the stock. Provided that we would always want to take a constant share of the stock (other strategies are possible and could also be better), we would always need all our boats. Each boat would always take a constant share of the stock, but catch less in years when there is a dearth of fish. It is also possible that we would want to vary the utilization of our fleet in an inverse proportion to the stock, using more boats (or operating each boat for a longer period of time) when the stock is small, and vice versa. This would, for example, be the case if we wanted to maintain a constant catch despite fluctuations in the harvestable biomass, provided that the catch per unit of effort is always proportional to the size of the stock. This will be discussed in greater detail in Chapter 4.

At the opposite extreme we have the case when the catch per unit of effort is

Table 2.1 Marginal catch and marginal revenue of a fishing fleet when allowable catches vary randomly between 0 and 0.9, with a uniform probability distribution.

Boats	Catch capacity	Marginal catch	Marginal revenue
0	0		
		0.09	0.9
1	0.1		
		0.08	0.8
2	0.2		
		0.07	0.7
3	0.3		
		0.06	0.6
4	0.4		
		0.05	0.5
5	0.5		
		0.04	0.4
6	0.6		
		0.03	0.3
7	0.7		
		0.02	0.2
8	0.8		
		0.01	0.1
9	0.9		
		0	0

constant and independent of the stock size. Here the fishing capacity of the fleet is well defined, depending only on the size of the fishing fleet. If the catch per year must be restricted because the fish stock is in a bad shape, the use of the fishing fleet must be restricted. Only if we always wanted to take a constant catch per year irrespective of the shape the stock is in would it be necessary always to use the fleet to its full capacity. There certainly are economic arguments in favour of such a strategy (see Chapter 4), but it might not be biologically feasible or advisable.

Since we do not always need all the boats, although they will cause some costs irrespective of whether they are used or not, the question arises as to the impact of the variability of fish stocks on the optimal fleet size. We already know that the relationship between the surplus growth of the stock and the size of the stock is an important determinant of the optimal effort or fleet size, but here we have another important factor, the variability of the stock growth for non-stock-dependent reasons. Here we shall use a simple example to demonstrate the importance of variability and ignore the stock–growth relationship. A more thorough treatment of the problem will be offered in Chapter 4.

Suppose that the catchable biomass depends to so high a degree on random effects that the relationship between the breeding stock and the exploitable biomass at a later date can be ignored, except that some safety level of breeding stock must be left behind each year. Let the total allowable catch vary in intervals of 0.1 over the interval 0–0.9 and assume that all these values are equally probable. Let the catch per boat (unit of effort used) be 0.1. Finally, let the unit price of fish net of short run operating cost be 10. Table 2.1 shows the additional catch per year of one additional boat. The reasoning behind the figures in the table is the following: in one of every 10 years the total allowable catch is likely to be zero, so the catch per year of just one boat will be $0.1 \times 0.9 = 0.09$. If we already have one boat, it will catch 0.1 at full capacity, and we will find use for it in 9 of every 10 years on the average. If we add one more boat, we will need it in only 8 of every 10 years on the average, because in 2 of every 10 years the total allowable catch will be only 0.1 or less. Hence the additional catch of an incremental boat (marginal catch) would only be $0.1 \times 0.8 = 0.08$ on the average if we already have one boat.

How many boats should be bought? That depends on the annual fixed cost of a boat. If it is higher than 0.9, not even the first boat will pay for itself, even if it lands a value equal to 1 per year when it is fully used, as it will not be used more than 9 years of every 10 on the average. If the annual fixed cost per boat is as low as 0.5, it makes sense to invest in five boats, as the last boat will exactly pay for itself and the previous four will yield a profit. Only if the fixed cost is as low as 0.1 would it make sense to invest in a fleet of nine boats, which is large enough to take the maximum total allowable catch. This illustrates how the optimal fleet capacity depends on the cost of durable equipment and the variability of the allowable catch.

The problem of finding the optimum fleet capacity for variable stocks, and other questions related to fluctuations in stocks, will be discussed in greater detail in Chapter 4.

2.8 Optimum exploitation of fish stocks and free-access equilibrium

In this section we derive mathematically the conditions for optimum exploitation of a fish stock over time. We then show how free-access equilibrium depends on biological and economic parameters, both in general and for special cases of the production function, the cost of effort function, and the demand function for fish. We also briefly consider possible development paths of a fishery and how likely it is that an equilibrium will be reached.

2.8.1 *The conditions for optimum exploitation*

Let the production function of the fishery be given by

$$Y = F(Z,S) \tag{2.4}$$

where Y is the catch of fish, Z is fishing effort, and S is the size of the fish stock.

Let the growth rate of the fish stock be specified as

$$dS/dt = G(S) - F(Z,S) \tag{2.5}$$

where $G(S)$ is the growth function of the stock.

Maximization of the present value of future profits in the fishery implies

$$\max \int_0^{\infty} [pF(Z,S) - C(Z)]e^{-rt}dt \tag{2.6}$$

where p is the price of fish, $C(Z)$ is the cost of fishing, and r is the discount rate. In a more general formulation, p, F, C and r could depend explicity on time. We will not consider that here, but the analysis is straightforward if more complicated. The price p is likely to depend on the catch volume, so instead of a constant p we would have a function $p(Y)$. This case will be considered below.

Problems such as (2.6) can be solved by applying the so-called maximum principle, which involves maximizing the so-called Hamiltonian function (H). This function consists of the maximand in Equation (2.6), to which one adds the function determining the rate of growth of the stock, multiplied by an auxiliary multiplier that changes over time (see, for example, Neher, 1990):

$$\max H(Z,S,t) = [pF(Z,S) - C(Z)]e^{-rt} + \lambda(t)[G(S) - F(Z,S)] \tag{2.7}$$

Necessary and sufficient conditions for the maximum of $H(Z,S,t)$, provided it is concave in Z and S, are as follows:

$$\partial H/\partial Z = [p(\partial F/\partial Z) - \partial C/\partial Z]e^{-rt} - \lambda(\partial F/\partial Z) = 0 \tag{2.8}$$

$$d\lambda/dt = -\partial H/\partial S = -p(\partial F/\partial S)e^{-rt} - \lambda(\partial G/\partial S - \partial F/\partial S) \tag{2.9}$$

Solving (2.8) for λ and differentiating with respect to time yields

$$d\lambda/dt = \{-r[p - (\partial C/\partial Z)/(\partial F/\partial Z)] + d[(\partial C/\partial Z)/(\partial F/\partial Z)]/dt\}e^{-rt} \tag{2.10}$$

For stationary solutions (that is, with constant effort and stock), the last term will be equal to zero. Setting (2.9) equal to (2.10) and substituting (from 2.8) for λ then yields

$$r - \partial G/\partial S = (\partial C/\partial Z)(\partial F/\partial S)/[p(\partial F/\partial Z) - \partial C/\partial Z] \tag{2.11}$$

Three special cases of this are discussed.

Case (1): $F(Z,S) = Z$

A mathematical problem with this case is that (2.7) becomes linear in Z. However, only (2.8) is compatible with sustained exploitation. As $\partial F/\partial S = 0$, Equation (2.11) becomes

$$r = \partial G/\partial S$$

which is Equation (2.2)

Case (2): $F(Z,S) = ZS$

A similar mathematical problem arises here; (2.7) is linear in Z, but only (2.8) is compatible with sustained exploitation.

Equation (2.11) becomes

$$r - \partial G/\partial S = (\partial C/\partial Z)Z/S[p - (\partial C/\partial Z)/S]$$

With $\partial C/\partial Z = c$, the cost per unit of fish caught, c_y, will be

$$c_y(S) = cZ/ZS = c/S$$

so that

$$\partial c_y/\partial S = -c/S^2$$

and (2.11) becomes

$$r - \partial G/\partial S = -(\partial c_y/\partial S)ZS/(p - c_y)$$

which is equivalent to Equation (2.3), noting that $ZS = G(S)$ in a stationary solution.

Case (3): Same as (2), but with $r = 0$

An additional mathematical problem here is that the integral in (2.6) does not converge, which violates the transversality conditions for the application of the maximum principle. However, we can proceed by maximizing the annual profit, noting that the steady annual flow of profit must in this case always prevail over transient profit in the long run. This involves

$$\max pG(S) - (c/S)G(S) \qquad (2.12)$$

where we have used the unit cost function derived under Case (2). The maximum condition is

$$\partial G/\partial S = -(c/S^2)G(S)/(p - c/S) = (\partial c_y/\partial S)G(S)/(p - c_y) \qquad (2.13)$$

which is the special form of (2.3) derived for $r = 0$ in Section 2.7.2.

The case $p(Y)$

In the case where price depends on the quantity caught, we have the demand function

$$p = p(Y) \qquad (2.14)$$

If we want to maximize profits, we substitute revenue, $R = p(Y)Y$ for pY in Equation (2.6). As argued in Section 2.5, maximization of the present value of profits is not appropriate in this case for deriving the optimum fishing effort, unless we are dealing with a country that exports fish. Instead we should maximize the difference between consumers' willingness to pay and the cost of fishing. Consumers' willingness to pay, or consumer surplus, for the quantity $\bar{Y}$ is

$$\int_0^{\bar{Y}} p(Y)dY \qquad (2.15)$$

Substituting (2.15) for pY in (2.6) gives

$$\max \int_0^{\infty} \left[\int_0^{Y} p(x)dx - C(Z) \right] e^{-rt} dt \qquad (2.6')$$

where $Y = F(Z,S)$

Substituting this in (2.7) and taking derivatives gives the following necessary conditions for maximum, in a stationary solution:

$$\partial H/\partial Z = [p^0(\partial F/\partial Z) - \partial C/\partial Z]e^{-rt} - \lambda(\partial F/\partial Z) = 0 \qquad (2.8')$$

$$d\lambda/dt = -\partial H/\partial S = -p^0(\partial F/\partial S)e^{-rt} - \lambda(\partial G/\partial S - \partial F/\partial S) \qquad (2.9')$$

where $p^0 = p(Y^0)$ and $Y^0 = F(Z^0,S^0)$, or the price for the marginal unit when the quantity supplied is determined by the optimal value of Z and S. Structurally this is exactly the same solution as above and could be obtained by maximizing the present value of profits at the constant price p^0, as asserted in Section 2.5.

2.8.2 *Free-access equilibrium*

Equilibrium in the fishery involves simultaneous biological and economic equilibrium. The biological equilibrium condition is that the rate of change of the stock be zero:

$$dS/dt = G(S) - F(Z,S) = 0 \qquad (2.16)$$

The economic equilibrium condition is that there be no change in effort. Profit is the driving force of effort; if the profit is positive and supposed to remain so for some time, this will entice people to invest in fishing capacity (effort), and vice versa. We shall simplify this to the point of assuming that the rate of change in effort is proportional to the profit of the marginal unit of effort. Measuring effort in homogeneous units, the catch per unit of effort will be the same for all units. The profit of the marginal unit is therefore

$$p(Y)F(Z,S)/Z - \partial C/\partial Z$$

From this we get

$$dZ/dt = a[p(Y)F(Z,S)/Z - \partial C/\partial Z] \qquad a > 0 \qquad (2.17)$$

which gives us the following equilibrium condition:

$$p(Y)F(Z,S)/Z - \partial C/\partial Z = 0 \qquad (2.18)$$

The equilibrium conditions (2.16) and (2.18) can be shown as curves in a diagram with Z and S on the axes. We shall refer to these as the biological and the economic equilibrium curve, respectively. The full equilibrium will be at the point of intersection of these curves. For later reference we need expressions for the slopes of these curves. Differentiating (2.16) yields

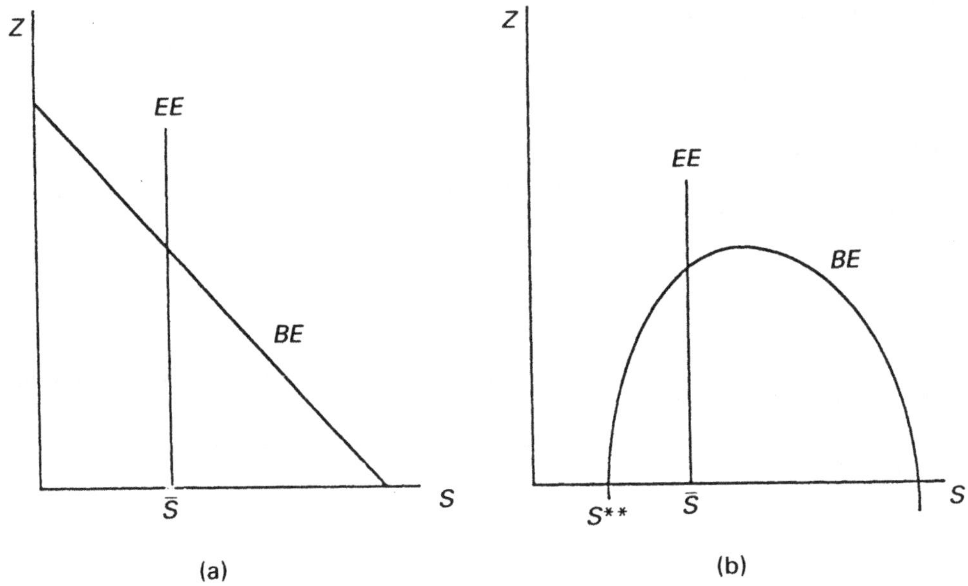

Fig. 2.13 Free-access equilibrium is at the intersection of the biological equilibrium curve (*BE*) and the economic equilibrium curve (*EE*). Catch per unit of effort is proportional to the size of the stock, price of fish and cost per unit of effort are constant. In (a) relative growth rate falls monotonically with increasing stock; in (b) relative growth rate increases with the stock (S) for 'small' S; a lowest viable stock $S^{**} > 0$.

$$\frac{dZ}{dS} = \frac{\partial G/\partial S - \partial F/\partial S}{\partial F/\partial Z} \tag{2.19}$$

while differentiating (2.18), making use of $Y = F(Z,S)$, gives

$$\frac{dZ}{dS} = \frac{p\dfrac{\partial F}{\partial S}\left[\dfrac{\partial p}{\partial Y}\dfrac{Y}{p} + 1\right]}{-p\dfrac{\partial F}{\partial Z}\left[\dfrac{\partial p}{\partial Y}\dfrac{Y}{p} + 1\right] + pY/Z + Z\dfrac{\partial^2 C}{\partial Z^2}} \tag{2.20}$$

We look now at some special cases.

Case (1): $F(Z,S) = ZS$, constant $\partial C/\partial Z$ and p

Equation (2.16) becomes

$$G(S) = ZS \tag{2.16.1}$$

and hence the biological equilibrium curve becomes

$$Z = G(S)/S$$

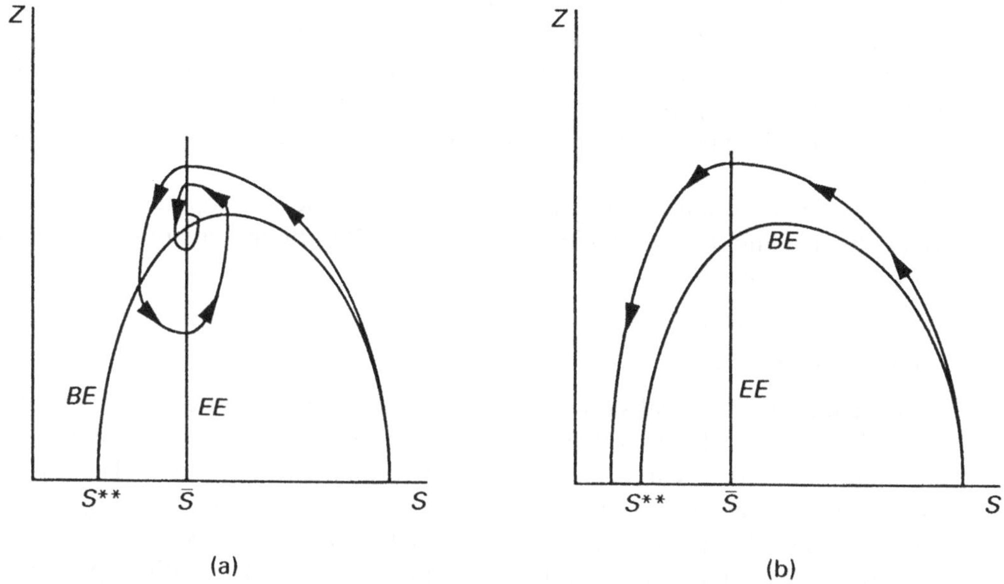

Fig. 2.14 Bioeconomic equilibrium: (a) development path converges on the equilibrium point; (b) development path overshoots the equilibrium and leads to extinction. *BE* is the biological equilibrium curve and *EE* is the economic equilibrium curve.

the slope of which is

$$dZ/dS = [\partial G/\partial S - G(S)/S]/S \tag{2.19.1}$$

If $\partial G/\partial S < G/S$ for all S, then $dZ/dS < 0$ for all S, as shown by the biological equilibrium curve '*BE*' in Fig. 2.13a. For stocks with a rising average growth rate (G/S), as can happen for 'small' S and does happen for stocks with a minimum viable size $S^{**} > 0$ (see Fig. 2.1), $\partial G/\partial S > G/S$ for small S, so that $dZ/dS > 0$ for small S, as shown by the curve '*BE*' in Fig. 2.13b. The curve in Fig. 2.13b implies $S^{**} > 0$

Consider then the economic equilibrium. This is given by Equation (2.18), which becomes

$$pS = \partial C/\partial Z \tag{2.18.1}$$

Given that $\partial C/\partial Z$ is constant, there is only one value of $S(\bar{S})$ that is compatible with economic equilibrium. The economic equilibrium curve therefore is a straight line, like the lines '*EE*' in Fig. 2.13 (checking the slope, we see that the denominator of Equation (2.20) is zero). Note that $\bar{S}$ is smaller the higher p or the lower $\partial C/\partial Z$.

The free-access equilibrium is at the point of intersection between the curve '*BE*' and the line '*EE*'. If '*EE*' is too far to the right, no such equilibrium exists.

In this case fishing is not profitable; the price is too low in relation to the cost of effort (the ratio $p/(\partial C/\partial Z)$ is too low). If '*EE*' is too far to the left, so that $\bar{S} < S^{**}$, no equilibrium exploitation is viable either. In this case the fishery is too profitable (the ratio $p/(\partial C/\partial Z)$ is too high), and the incentive to expand fishing effort will not disappear until the stock has been depleted below its lowest viable level.

Will the equilibrium be attained? We can study the development path of the fishery by dividing the rate of change in effort (dZ/dt) by the rate of change of the stock (dS/dt) to get

$$\frac{dZ}{dS} = \frac{a(pS - \partial C/\partial Z)}{G(S) - ZS}$$

This path has a vertical slope when it crosses the biological equilibrium curve, as $dS/dt = 0$ everywhere on that curve, and a horizontal slope when it crosses the economic equilibrium curve, as $dZ/dt = 0$ everywhere on that curve. There is one such path from each initial situation, and no two paths can ever cross. Figure 2.14 shows two possible paths from an initial situation with an unexploited stock, for the biological equilibrium function shown in Fig. 2.13b. The path in part (a) of the figure spirals in towards the equilibrium point, while the path in part (b) never reaches it. The path in part (b) illustrates the important point that a fish stock might become extinct as a result of exploitation even if a free-access equilibrium exists; the fishery might simply fail to achieve it.

Whether or not the fishery will attain the equilibrium depends on the laws of motion of the fishery, in particular how quickly the effort expands or contracts in response to changes in profitability, and the exact location of the equilibrium. The further to the left the equilibrium point is, the less likely it is that it will be reached.

Consider a simple example. Let the growth function be specified as

$$G(S) = (1 - S)(S - 0.1) \tag{2.21}$$

implying that the natural equilibrium of the stock is 1 and the minimum viable size is 0.1. The rate of change of the stock is

$$dS/dt = (1 - S)(S - 0.1) - ZS$$

According to Equation (2.17), the rate of change of effort is

$$dZ/dt = a(pS - \partial C/\partial Z)$$

Now, by increasing time in small steps, we may approximate the development of effort and stock over time as follows. Over the time interval Δt the effort will change to

$$Z_{t+\Delta t} = Z_t + a(pS - \partial C/\partial Z)\Delta t$$

Applying this effort, the stock changes to

$$S_{t+\Delta t} = S_t + [(1 - S_t)(S_t - 0.1) - Z_{t+\Delta t}S_t]\Delta t$$

Figures 2.15 and 2.16 trace the development of effort and stock over time, for $a = 0.01$, $p = 1$, and two alternative values of $\partial C/\partial Z$. For $\partial C/\partial Z = 0.5$, the equilibrium stock is 0.5 and the equilibrium effort is 0.4. As parts (a) and (b) of Fig. 2.15 show, both effort and stock move in cycles after starting from a level of no exploitation, but the cycles dampen and the equilibrium is reached asymptotically.

For the 'low' cost of $\partial C/\partial Z = 0.2$, the equilibrium stock is 0.2 while the equilibrium effort again is 0.4. This equilibrium will not be reached. As Fig. 2.16 shows, the effort will not be reduced quickly enough and the stock will be driven down to zero. Different specifications of the effort dynamics would allow the equilibrium to be reached; an example is letting the contraction of effort occur more rapidly once the stock has been driven down below its equilibrium level.

This last remark indicates that our little example may be too optimistic with respect to how likely free-access fisheries are to reach an equilibrium. Quite often it would be more realistic to suppose that effort changes more slowly in response to losses than to profits. The owners of vessels that have no alternative use will be happy to recover some of the fixed cost and will not stop operating until the revenue falls below the operating cost or the boats wear out. On the other hand, those who invest in new vessels will do so only if they expect to recover all costs. Thus higher profits are needed to provide an incentive to expand a fishing fleet than to keep it employed once capital has been invested in boats with little or no alternative use.

Case (2): $F(Z,S) = ZS$, constant p but rising $\partial C/\partial Z$

This does not affect the biological equilibrium curve (*BE*), but the economic equilibrium curve (*EE*) bends to the right, as (from 2.20) its slope is

$$\frac{dZ}{dS} = \frac{p}{\partial^2 C/\partial Z^2} > 0 \qquad \text{since } \partial^2 C/\partial Z^2 > 0 \tag{2.20.2}$$

In this case it is possible to have two equilibria with the '*BE*' curve shown in Figs. 2.13b and 2.14b. This possibility is illustrated in Fig. 2.17. One of these equilibria (the one closer to the origin) is unstable, as any small movement from that point, should it ever be attained, will send the stock moving either to the other equilibrium or to extinction. This can be seen as follows. Suppose effort changes so quickly that the stock is always at a level compatible with economic equilibrium, that is, on the curve '*EE*'. Below the curve '*BE*' it will be increasing, and above that curve it will be decreasing.

The positive slope of the economic equilibrium curve increases the chance that a free-access equilibrium will exist and that the fishery may come to rest there. The common-sense interpretation of this is that the increasing unit cost of effort will make it less attractive than otherwise to expand effort to levels that are high enough to wipe out the stock.

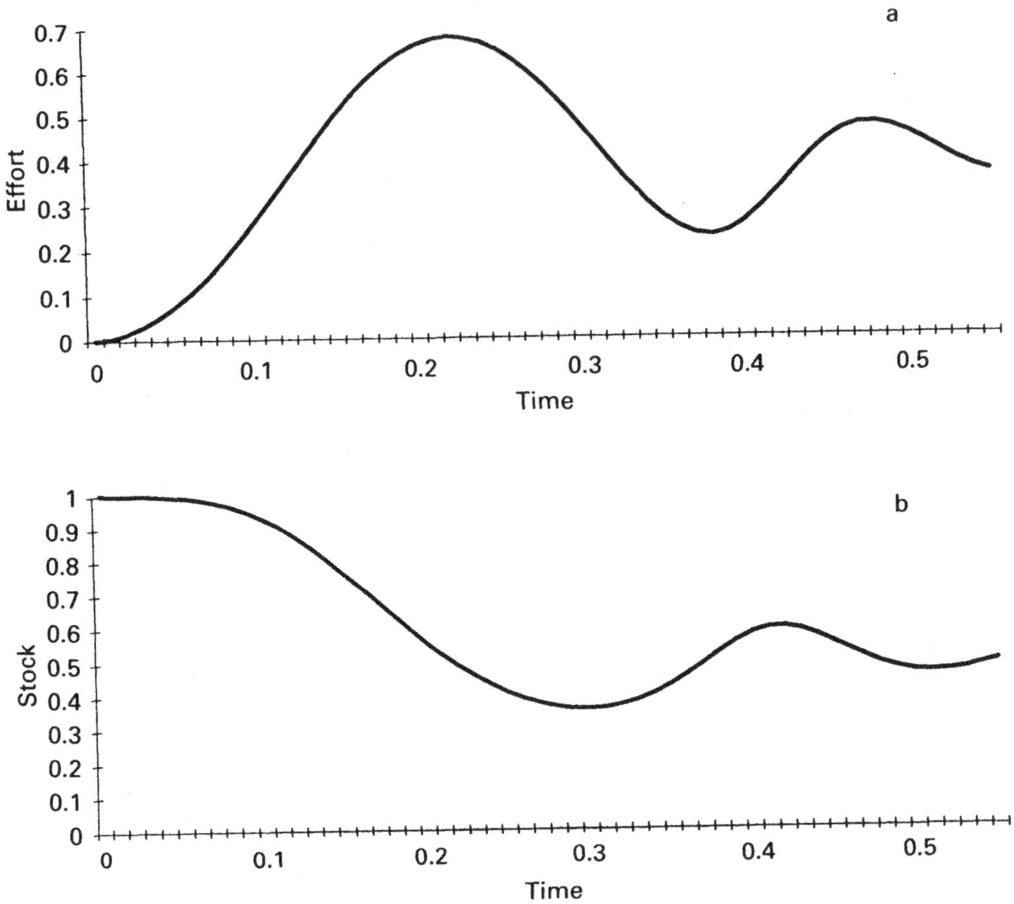

Fig. 2.15 Development of a fishery starting in a natural equilibrium, with effort changing in response to profit. Cost of effort is 0.5.

The biological equilibrium curve in Fig. 2.17 was generated with the growth curve (2.21), while the cost of effort function was specified as $C = Z^2$.

Case (3): $F(Z,S) = ZS$, $\partial C/\partial Z$ constant but $\partial p/\partial Y < 0$

In this case the slope of the economic equilibrium curve becomes

$$\frac{dZ}{dS} = -\frac{pZ\left[\dfrac{\partial p}{\partial Y}\dfrac{Y}{p} + 1\right]}{SY(\partial p/\partial Y)} = -\frac{pZ(1 - 1/\eta)}{SY(\partial p/\partial Y)} \tag{2.20.3}$$

where $\eta = -(\partial Y/\partial p)p/Y$, which is called the price elasticity of demand. If demand is elastic ($\eta > 1$), $dZ/dS > 0$ as in Case (2), since $\partial p/\partial Y < 0$. If, on the other hand, demand is inelastic ($\eta < 1$), then $dZ/dS < 0$. This is shown in Fig.

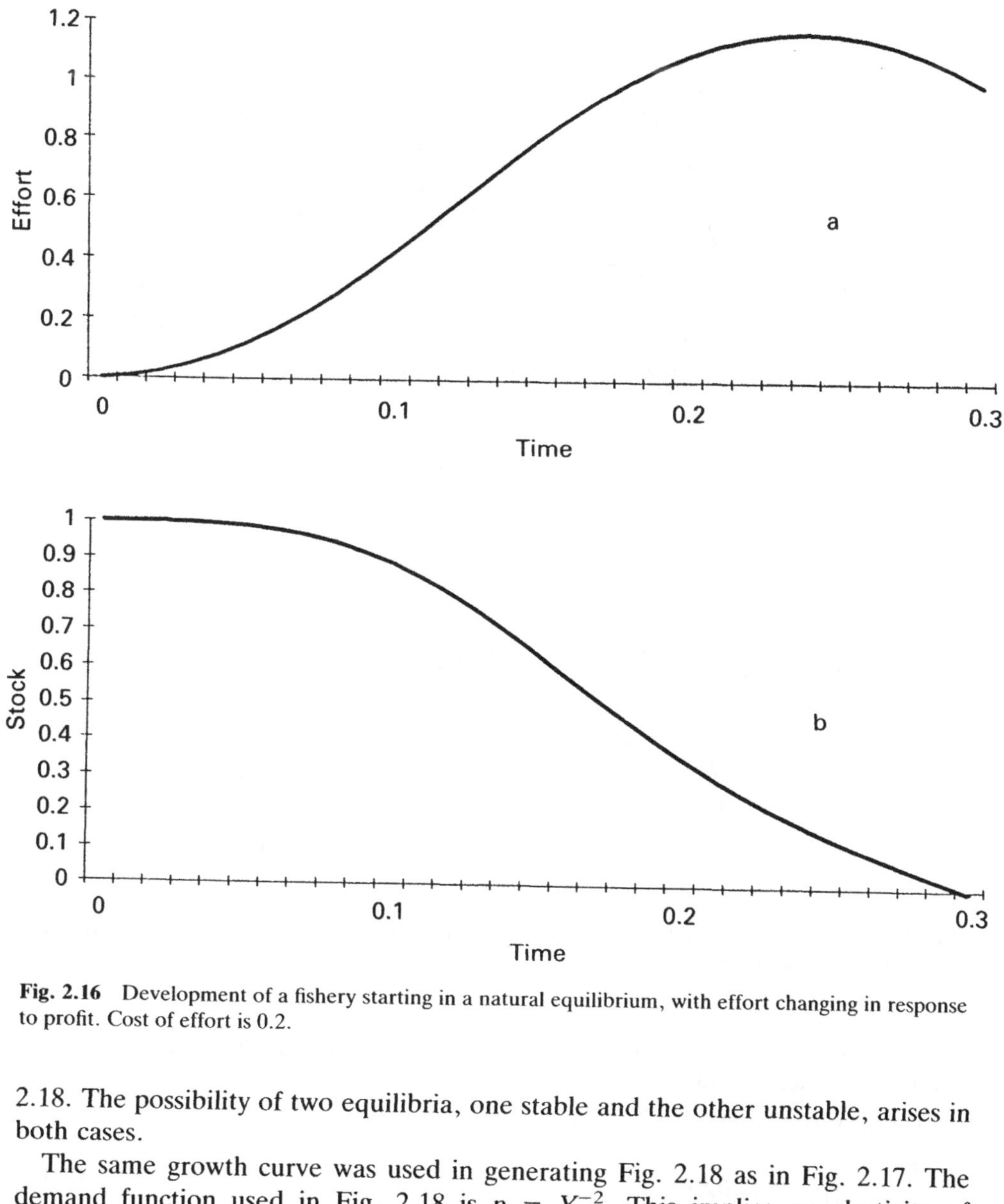

Fig. 2.16 Development of a fishery starting in a natural equilibrium, with effort changing in response to profit. Cost of effort is 0.2.

2.18. The possibility of two equilibria, one stable and the other unstable, arises in both cases.

The same growth curve was used in generating Fig. 2.18 as in Fig. 2.17. The demand function used in Fig. 2.18 is $p = Y^{-2}$. This implies an elasticity of demand of −0.5. Using $p = 1/\sqrt{Y}$, which implies an elasticity of demand of −2, produces the same result as Fig. 2.17.

Case (4): $F(Z,S) = Z$, p and $\partial C/\partial Z$ constant

Here the biological equilibrium condition becomes

$$G(S) = Z \tag{2.16.4}$$

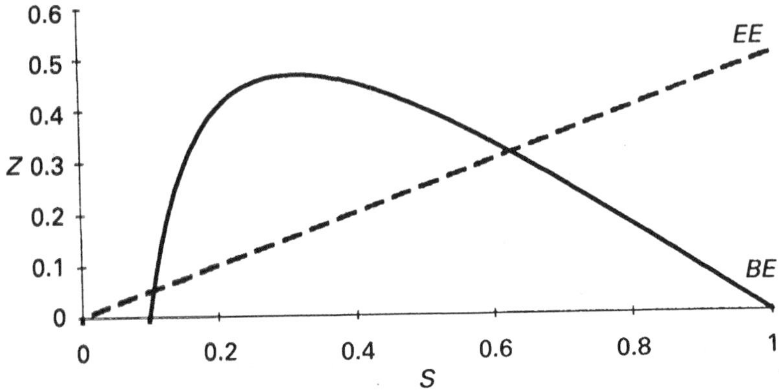

Fig. 2.17 Bioeconomic equilibrium. Rising marginal cost of effort or elastic demand. *BE* is the biological equilibrium curve; *EE* is the economic equilibrium curve.

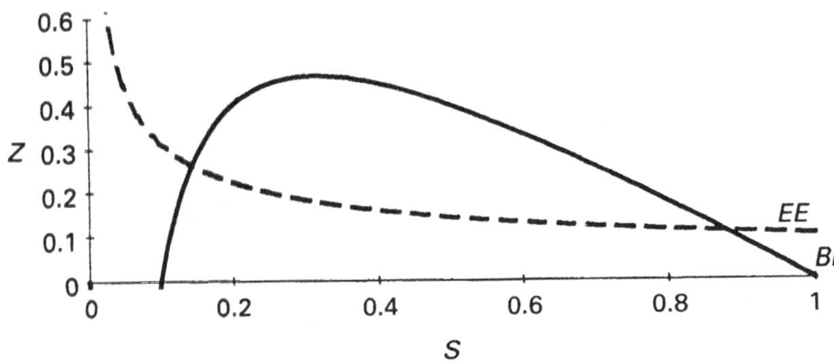

Fig. 2.18 Bioeconomic equilibrium. Inelastic demand. *BE* is the biological equilibrium curve; *EE* is the economic equilibrium curve.

and the biological equilibrium curve becomes identical with the growth curve $G(S)$, as shown in Fig. 2.19.

The rate of change of effort is

$$\frac{dZ}{dt} = a[p - \partial C/\partial Z] \qquad (2.17.4)$$

There is no way that this will be zero except by coincidence. Either $p > \partial C/\partial Z$ and the effort will expand without limit until the fish stock has been depleted, or $p < \partial C/\partial Z$ and the fishery will never be profitable.

Case (5): $F(Z,S) = Z$, rising $\partial C/\partial Z$ or $\partial p/\partial Y < 0$

If the price depends on the quantity taken or the marginal cost of effort depends

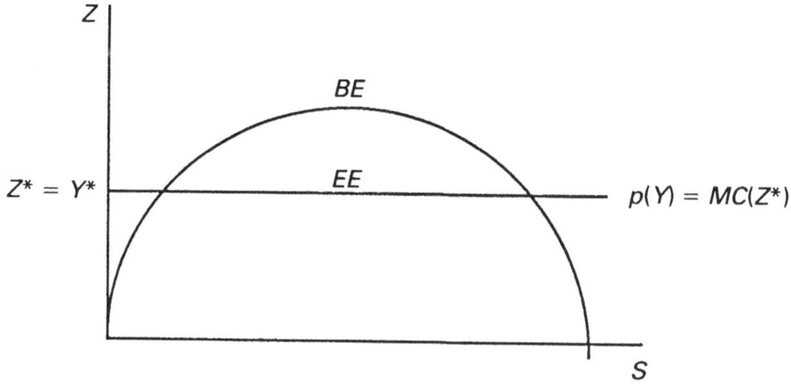

Fig. 2.19 Bioeconomic equilibrium. Constant catch per unit of effort. Falling price as landings increase or increasing marginal cost of effort as effort increases. *BE* is the biological equilibrium curve; *EE* is the economic equilibrium curve.

on the level of effort, it is possible that $dZ/dt = 0$ for some level of landings or effort, respectively. The economic equilibrium condition is then given by

$$p(Y) = \frac{\partial C}{\partial Z}(Z)$$

If p falls sufficiently fast with increasing Y or $\partial C/\partial Z$ increases sufficiently fast with increasing Z, it is possible that the economic equilibrium line will intersect the biological equilibrium curve, as shown in Fig. 2.19. Two free-access equilibria will then exist, one stable and the other unstable, as in some of the previous cases. The common-sense interpretation of this is that an unsustainable expansion of the fishery may not be profitable if the marginal profit falls as the fishery is expanded, either because the price falls with increased landings or the marginal cost of effort increases with effort.

Case (6): $\partial^2 F/\partial S^2 < 0$ (a weakened stock effect)

The biological equilibrium condition is

$$G(S) - F(Z,S) = 0 \tag{2.16}$$

The biological equilibrium curve in Fig. 2.20 was produced by the same growth function as used in Figs 2.17 and 2.18, but with the production function specified as $Y = Z\sqrt{S}$. This specification of the production function can be said to imply 'halving' of the stock effect; that is, the impact of S on the catch per unit of effort. We see that this 'deflates' as it were the biological equilibrium curve and makes a point of intersection with the economic equilibrium curve less likely to occur. The common-sense interpretation of this is that the less the catch per unit of effort falls as the stock is depleted, the less the profitability of the fishery will be eroded

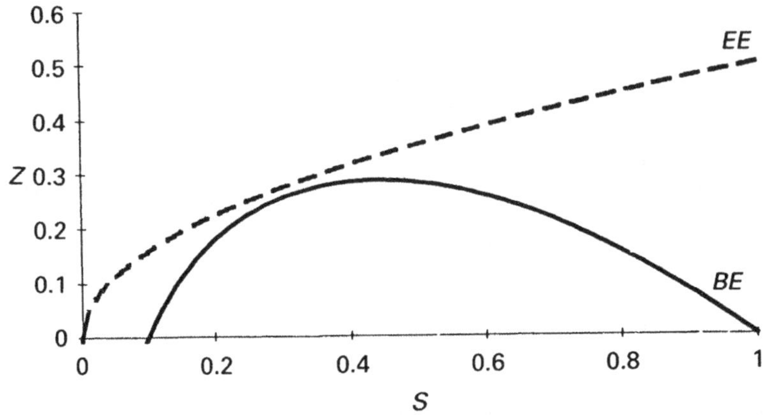

Fig. 2.20 Bioeconomic equilibrium. 'Halving' of stock effect. *BE* is the biological equilibrium curve and *EE* is the economic equilibrium curve.

by depleting the stock, and the less likely it is that the profit will be zero at any viable stock level.

For the specific growth function used here and the cost function used in Fig. 2.17 ($C = Z^2$), the economic equilibrium curve is

$$p\sqrt{S} = 2Z \tag{2.18.6}$$

and its slope is

$$\frac{dZ}{dS} = p/4\sqrt{S} \tag{2.20.6}$$

The economic equilibrium curve is also shown in Fig. 2.20. It is 'inflated', as it were, with fixed end points at $S = 1$ and $S = 0$. The intersection of the two equilibrium curves, '*BE*' and '*EE*', is now less likely, and in fact they do not intersect at all in Fig. 2.20. This demonstrates how a weakened stock effect makes the existence of a free-access equilibrium less likely.

Chapter 3
Applied Fishery Models

3.1 General remarks

Qualitative answers like 'optimal fishing effort is less than free-access effort' do not provide very precise targets for the management of fisheries. To determine exactly how much should be caught, how many boats should be built, and how much effort should be used in any given period, we need a numerical model of the interaction between the fishing fleet, the catch it takes, the size and growth of the fish stock, etc. The model must be in a form that is suitable for mathematical manipulation or calculation on a computer.

At this point a word of caution is needed. The answers provided by numerical models depend on their mathematical form and the values of their parameters. A change in the specification of a model, or even a slight modification of the values of its parameters, may radically change the numerical answers the model provides. This is a serious problem. The values of the parameters in a model are typically not very precisely known, and the specification of the model is a matter of approximating complex and imperfectly known processes of nature with mathematical equations that are simple to manipulate. A good precautionary rule is to try alternative model specifications and parameter values in order to test the robustness of the results. In some cases it may be possible to use statistical procedures to provide a numerical evaluation of the validity of the results in the form of confidence intervals around the precise numerical answers provided by the models.

The simplest applied fishery models are based on a stock–growth relationship of the kind discussed in Chapter 2. The mathematical functions written in a general form in Chapter 2 are given a specific form suitable for numerical calculations, and their parameters are estimated from available data. One way of classifying such models is by whether changes over time are supposed to take place continuously or at certain intervals. In continuous time models fishing is steadily going on, the fish grow without interruption, albeit at a declining rate, and fish disappear from the stock continuously. In discrete time models changes take place at regular intervals. The growth that has occurred over a period (a month or a year or whatever) is added to the stock, and the catch that has been taken is subtracted.

To some extent the choice between the two types of models is a matter of computational and mathematical simplicity. Discrete models are typically better for computational purposes, while continuous models usually are simpler to manipulate with analytical techniques such as calculus. Besides mathematical convenience, it is important to approach the processes of nature as closely as possible. Even if the processes of nature are continuous, some of them have a distinct seasonal pattern and are most appropriately modelled in discrete time. In temperate waters with clear seasonal changes in temperature, the growth of fish and recruitment of young cohorts are seasonal, and sometimes the fishing is seasonal as well because of seasonal migrations or variations in the weather. In warmer waters these processes typically are more or less continuous; the fish are steadily growing, spawning and recruitment take place continuously, and so does fishing. But seasonality in fishing and the recruitment of fish can also occur in tropical waters, because of seasonality in the weather (rainfall, for example).

Greater realism and reliability can be obtained by moving beyond the simple stock–growth relationship. Age-structured models are an important step in this direction. When the reproduction of fish is seasonal, this gives rise to distinct cohorts of fish. The most common seasonality probably is the annual cycle, with spawning and recruitment taking place over a relatively short period once every year. Such cohorts are therefore referred to as year classes. Age-structured models follow each cohort of fish through time, and stock growth and catch are explained by the growth of fish in each cohort and removals of fish. This approach is extensively used in the analysis of fisheries in temperate waters where fish stocks are composed of distinct year classes.

In this chapter we shall discuss three applied fishery models extensively used by fisheries biologists. We begin with the Schaefer model, which is a simple stock–growth model of the continuous type. We then move on to the Ricker model, which is a discrete time stock–growth model. Lastly we discuss the Beverton–Holt year class model. This model distinguishes between different age groups, or year classes, of fish and follows each year class through its life history as affected by fishing, natural mortality, and growth of individual fish. A relationship between growth and (equilibrium) stock size is implicit rather than explicit in this model. The emphasis will be on how these models can be extended to include economic analysis rather than on a detailed biological or mathematical exposition.

3.2 The Schaefer model

This model was developed by the American fisheries biologist M.B. Schaefer (Schaefer 1954, 1957). In the Schaefer model the rate of surplus growth is specified as

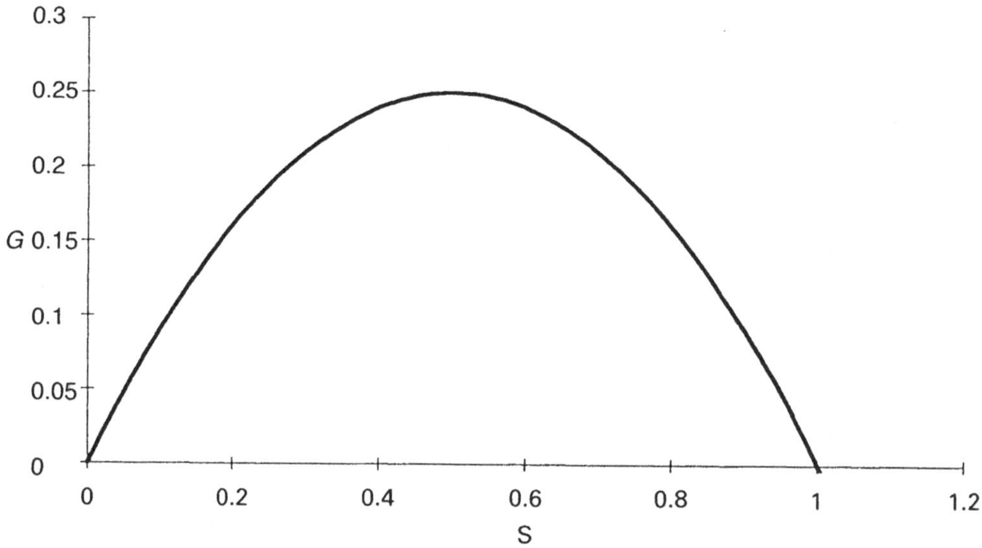

Fig. 3.1 Growth function: $G = S(1 - S)$

$$G(S) = aS(1 - S/k) \tag{3.1}$$

This implies a parabolic growth curve, as shown in Fig. 3.1. In natural equilibrium the surplus growth is zero, which implies that $S = k$, so that k can be interpreted as the carrying capacity of the environment. The surplus growth curve is symmetric and rises at a decreasing rate to a maximum at $k/2$, intersecting the x-axis at $S = k$, the natural equilibrium level of the stock. This can be seen mathematically by calculating the first derivative of (3.1):

$$G'(S) = a(1 - 2S/k) \tag{3.2}$$

If fishing effort always removes a constant proportion of the stock and the catch is always equal to the surplus growth, this model can be transformed in a very simple way to a catch–effort model. Before we do so, let it be emphasized that these are very strong assumptions. As discussed in Chapter 2, the catch per unit of effort directly aimed at catching fish will be proportional to the stock size only if the density of fish is proportional to the size of the stock. The assumption that the catch will always be equal to the surplus growth is also restrictive. Consider a change in fishing effort. This will initially involve an adjustment of the stock; the stock will increase if the effort is diminished and vice versa. Only after the initial adjustment period is over will the catch be equal to the surplus growth. The adjustment period will be short, however, for stocks with a high turnover rate, such as fast-growing and short-lived species typically living in warm waters.

Given the said assumptions, how can the model be transformed into a catch–

effort model? With the catch per unit of effort being proportional to the stock, we can write

$$Y = ZqS \tag{3.3}$$

where Y is catch and q is a factor of proportionality. Setting $Y = G(S)$ gives

$$ZqS = aS(1 - S/k)$$

which may be written

$$S = k(1 - Zq/a) \tag{3.4}$$

Using (3.4) in (3.3) gives

$$Y = Zqk(1 - Zq/a)$$

which may be written

$$Y = \alpha Z - \beta Z^2; \qquad \alpha = qk, \quad \beta = q^2k/a \tag{3.5}$$

Because q, k and a are constant parameters, α and β are also constant parameters. Dividing through by Z we get the catch per unit of effort as a linear function of effort. This gets us quite far, because data on fish catches and effort are usually available if anything at all is known about a fishery. Such data allow us to estimate the parameters α and β by linear regression of the catch per unit of effort on effort.

Having found the parameters of the yield–effort curve, we need to know how the price varies with the quantity landed and how the cost of fishing varies with fishing effort in order to find the economic optimum. For this it is necessary to have data on prices and landings and on the cost of the fishing boats. The best way to proceed in the analysis of cost is probably to survey a sample of representative boats. This would allow one to estimate the marginal cost of effort; that is, the cost of adding on a new boat, or of using one of the boats that exist. Note the distinction between short-term and long-term costs. If the fishing fleet is already too large, the question is how much of its capacity to use. In that case one would disregard costs that are fixed in the short term; that is, depreciation of durable equipment and the interest on the capital that is tied up in this equipment (capital costs). As old boats are gradually scrapped, the size of the fishing fleet approaches the optimal long-term size. Having reached the point at which worn out boats have to be replaced, the long-term cost becomes relevant.

As an illustration, consider the hypothetical example in Table 3.1. By linear regression we can estimate the coefficients α and β to

$$\alpha = 0.61227, \ \beta = -0.00057$$

from which we get

$$Y/Z = 0.61227 - 0.00057Z \tag{3.6}$$

$$Y = 0.61227Z - 0.00057Z^2 \tag{3.5'}$$

Table 3.1 Hypothetical example of catch and effort observed over eleven years.

Year	Catch	Effort	Catch/Effort
1	60	100	0.6
2	50	100	0.5
3	125	250	0.5
4	120	300	0.4
5	180	450	0.4
6	150	500	0.3
7	180	600	0.3
8	140	700	0.2
9	160	800	0.2
10	80	800	0.1
11	90	900	0.1

3.2.1 *Optimum yield*

Suppose the price (p) of fish is 10 and the long-term cost per unit of effort (c) is 1. Ignoring discounting and present values for the moment, the long-term economic optimum is where the marginal sustainable yield is equal in value to the cost of an additional unit of effort. By calculating the first derivative of the sustainable yield function (3.5′) we find the marginal sustainable yield:

$$dY/dZ = 0.61227 - 0.00114Z \tag{3.7}$$

the value of which is

$$p(dY/dZ) = 6.1227 - 0.0114Z \tag{3.7'}$$

Setting $p(dY/dZ) = 1$, the marginal cost of effort, gives optimum sustained effort $Z° = 449$ and optimum sustainable yield $Y° = 160$. This would give a sustained profit of 1151, which is more than 70 per cent of the total catch value. It may be noted that the optimum sustainable yield is slightly less than the maximum sustainable yield, and the optimum sustained effort considerably less than needed to take the maximum sustainable yield. These we may find by setting (3.7′) equal to zero and using (3.5′), which gives $Z_{msy} = 537$ and $Y_{max} = 164$.

According to the data, the actual effort in the last year on record is twice as great as the optimum effort, and also much greater than needed to take the maximum sustainable yield. How could this have happened? To find the equilibrium effort under free access we set the value of the catch per unit of effort equal to the cost per unit of effort, as argued in Chapter 2. This gives (see Equation (3.6))

$$p(Y/Z) = 6.1227 - 0.0057Z = 1$$

which implies $Z^* = 899$. The free-access fishery in this example is certainly wasteful; the sustained catch is only 90, which is only slightly more than one-half of the maximum sustainable catch, with no profit being obtained. The free-access effort is twice as great as the economically optimum effort, while the maximum sustainable yield effort is about 60 per cent of the free-access effort. Some 40 per cent of the free-access effort thus contribute nothing at all to the fishery, indeed this excessive effort is unequivocally wasteful in two ways; it causes unnecessary costs and results in a smaller catch than need be.

Given that the fishery has expanded beyond the economically optimal point, it would be appropriate to consider short-term costs only. Some units of effort (boats) are redundant, but if they cannot be used for any other purpose (such as other types of fishing), their capital cost cannot be recovered anyway, so the only relevant costs are operating costs such as fuel, labour costs, etc. Suppose the capital costs are one-half of the total cost per boat. This gives us a short-term cost of only 0.5. Setting $p(dY/dZ) = 0.5$ gives $Z^o = 493$. This is a bit more than the long-term optimum effort (449), but much less than the equilibrium effort under free access. The right policy in this case would be to decommission $900 - 493 = 407$ units of effort immediately, but continue using the remaining ones until we are left with the long-term optimum (449) after some of the 493 units have been worn out. At that point worn out boats will need to be replaced, and capital costs become relevant.

How about the people working on the boats that are not needed? There will be cases when there is no alternative employment for these people, in the short term at any rate. Here one must distinguish between the social or distributional reasons for having people employed in fishing and the efficiency reason for the same. Employing people on boats that contribute nothing to the value produced in the fishery is pure waste, and employing people on boats that reduce the value produced in the fishery, as in the example above, is even worse. This does not mean that these people should be left without any means to support themselves; no civilized society would accept that kind of policy. There are two ways to deal with this. One is to provide these people with an income regardless of whether they work or not. After all society can 'afford' this, because the net value produced by the fishery will, after an adjustment period of lower catches, be at least as great as it was when they were employed in fishing. The other method is to devise some work-sharing arrangement by which the cut-back in effort is shared by those who used to work full time in the fishery. This method might, for social reasons and reasons of keeping people in the labour force, be preferable to income support. A major challenge, however, is how to devise such schemes without destroying the incentive for people to seek a more productive employment. The gain from reducing employment in an industry such as an overexploited fishery, where a marginal unit of labour contributes nothing to the value produced, is greatest when people can be transferred to another industry where they will make a positive contribution. Such transfers can be expected to occur voluntarily if people can make a personal gain from doing so.

3.2.2 *The impact of discounting*

How does discounting of the future affect the results derived in this example? We can find this by an approach similar to the one we took in Chapter 2, but since we do not know the parameters of the growth function we have to rely on the yield–effort function instead. The question to be considered is this. Would it be desirable to expand effort further than to $Z = 449$, the value that maximizes sustainable annual profit? If we did so, our immediate gain would be approximately equal to the value of the catch per unit of effort minus the cost of a unit of effort, multiplied by the 'small' change in effort, $[p(Y/Z) - c]\Delta Z$. The present value of the long term loss would be the loss in sustainable profit caused by the 'small' change in effort divided by the rate of discount, $[p(dY/dZ) - c]\Delta Z/r$, a negative number indicating a loss. Adding the long- and short-term effects and observing that in optimum these must net to zero gives

$$\{p(Y/Z) - c + [p(dY/dZ) - c]/r\}\Delta Z = 0 \tag{3.8}$$

Ignoring costs for the moment, we see that the economic optimum would be on the right side of the maximum of the sustainable yield curve (dY/dZ would have to be negative). This implies a greater effort than needed to take the maximum sustainable yield and a smaller standing stock than supports the maximum surplus growth, which agrees with our conclusion in Chapter 2. Furthermore, we see that if the discount rate becomes very high and the long-term effect can be ignored, the free-access equilibrium where the value of the catch per unit of effort is equal to the cost per unit of effort becomes optimal. Hence it is possible to say that future catches are implicitly ignored under free access.

Using the parameter values we derived above and assuming a discount rate of 0.1 in Equation (3.8) gives

$$6.1227 - 0.0057Z^{o} - 1 + 61.227 - 0.114Z^{o} - 10 = 0 \tag{3.8'}$$

or $Z^{o} = 471$. This is slightly more than in the absence of discounting (449) but less than the maximum sustainable yield effort (537). If the fishing costs could be ignored altogether, we would get an optimum effort of 563, which is slightly greater than the maximum sustainable yield effort.

If the parameters of the Schaefer model (a and k) and the size of the fish stock at each particular time can be estimated, it might be possible to estimate a production function for the fishery; that is, a relationship between fishing effort, fish catch and the size of the exploited stock. The definition and measurement of effort and the specification of the production function pose formidable problems which will not be discussed here in any detail. A simple, but not unrealistic, way of specifying the production function is the multiplicative form

$$Y = \gamma Z^{\alpha} S^{\beta} \tag{3.9}$$

where γ, α and β are constant parameters. Taking logarithms permits one to use linear regression to estimate the parameters of the equation:

$$\ln Y = \ln\gamma + \alpha\ln Z + \beta\ln S$$

If both α and β are equal to 1, we are back with the case considered above, the realism of which was discussed in Chapter 2. The values of α and β reflect important characteristics of the fishery. Note that the catch per unit of effort as a proportion of the exploited stock is

$$Y/ZS = \gamma Z^{\alpha-1}S^{\beta-1}$$

If $\beta < 1$, this means that each unit of effort catches a greater proportion of the stock as the stock diminishes. This would occur if the density of the stock is not constant but increases as the stock declines, because it spreads itself over a smaller and smaller area. With $\beta = 0$ we obtain the case in which the catch per unit of effort is independent of the stock size and constant, provided $\alpha = 1$. If $\alpha < 1$, this means that each unit of effort catches less and less from a given stock the more effort is applied, which could be due to crowding effects.

Having estimated a growth function and a production function, it would be possible to use Equation (2.3) to calculate the optimum stock, catch and effort. Equation (2.3) is reproduced below as

$$[r - G'(S^o)][p - c_y(S^o)] = -c_y'(S^o)G(S^o) \tag{3.10}$$

We can use Equations (3.1) and (3.2) to express $G(S)$ and $G'(S)$, while the cost per unit of fish caught can be derived using the production function (3.9). The total cost of fishing is cZ, c being the cost per unit of effort. The catch is $Y = \gamma Z^\alpha S^\beta$, so the cost per unit of fish caught is

$$c_y(S,Z) = c\gamma^{-1}Z^{1-\alpha}S^{-\beta} \tag{3.11}$$

We can eliminate Z from this relation by observing that, in equilibrium, $Y = G(S)$. Setting (3.1) equal to (3.9) gives

$$\gamma Z^\alpha S^\beta = aS(1 - S/k)$$

Solving for Z we get

$$Z = \gamma^{-\frac{1}{\alpha}} S^{\frac{1-\beta}{\alpha}} \left[a\left(1 - \frac{S}{k}\right)\right]^{\frac{1}{\alpha}}$$

which when substituted in (3.11) gives

$$c_y(S) = c\gamma^{-\frac{1}{\alpha}} S^{\frac{1-\alpha-\beta}{\alpha}} \left[a\left(1 - \frac{S}{k}\right)\right]^{\frac{1-\alpha}{\alpha}} \tag{3.11'}$$

Taking the first derivative of this expression gives

$$c_y'(S) = c\gamma^{-\frac{1}{\alpha}} \left[a\left(1 - \frac{S}{k}\right)\right]^{\frac{1-\alpha}{\alpha}} S^{\frac{1-\alpha-\beta}{\alpha}} \left\{\frac{(1-\alpha-\beta)}{\alpha S} - \frac{1-\alpha}{\alpha k\left(1 - \frac{S}{k}\right)}\right\} \tag{3.12}$$

Let us look at two special cases of this, $\beta = 1$ and $\beta = 0$, with $\alpha = 1$. These special cases were discussed extensively in Chapter 2. Here we will demonstrate that the difference between optimum stock and effort levels in these two cases may be very significant.

In the two special cases, Equations (3.11′) and (3.12) become

$\beta = 1$:

$$c_y(S) = c\gamma^{-1}S^{-1}, \quad c_y'(S) = -c\gamma^{-1}S^{-2}$$

$\beta = 0$:

$$c_y(S) = c\gamma^{-1}, \quad c_y'(S) = 0$$

Inserting (3.1), (3.2), (3.11′) and (3.12) in (3.10) in the two cases gives:

$\beta = 1$:

$$(r - a + 2aS^{o}/k)(p - c/\gamma S^{o}) = ca(1 - S^{o}/k)/\gamma S^{o}$$

$\beta = 0$:

$$(r - a + 2aS^{o}/k)(p - c/\gamma) = 0$$

Solving for the optimal value of S in the two cases gives

$\beta = 1$:

$$S^{o} = (k/4a)\{(a - r + ac/\gamma kp) \pm [(a - r + ac/\gamma kp)^2 + 8arc/\gamma kp]^{1/2}\} \tag{3.13a}$$

$\beta = 0$:

$$S^{o} = (a - r)k/2a \tag{3.13b}$$

As stated in Chapter 2, the optimum stock size is independent of the price of fish and the unit cost of fishing when the price is independent of the quantity landed and the unit cost (c_y) is constant and independent of the size of the stock. This is the case $\beta = 0$.

We illustrate with estimates based on data of the Antarctic blue whale (Clark 1973, after J.A. Gulland). One set of estimates of the parameters of the Schaefer model is $a = 0.1$, $k = 210\,000$. Exploitation of the blue whale was stopped when only 5000 animals were believed to be left, so assuming that the price is equal to the unit cost at $S = 10\,000$ in the stock-dependent case is not unrealistic and probably on the high side. For a constant price, p, of whales, this gives, for the case when the unit cost depends on the stock:

$$p = c_y(S = 10\,000) = c/10\,000\gamma$$

which implies $c/\gamma p = 10\,000$. Table 3.2 shows how the optimal stock size and sustainable yield vary with the rate of discount in the two above cases.

This example illustrates several important points, some of which have already been discussed. Firstly, the impact of the discount rate on the optimum stock and sustainable yield is seen to be formidable; optimum extinction becomes a

Table 3.2 Optimum stock of the Antarctic blue whale as a function of the discount rate (r). $\beta = 1$: catch per unit of effort proportional to the stock size. $\beta = 0$: catch per unit of effort constant.

	$\beta = 1$		$\beta = 0$	
r	S^o	Y^o	S^o	Y^o
0	110 000	5238	105 000	5250
0.01	109 100	5242	94 500	5198
0.03	82 300	5005	73 500	4778
0.05	65 500	4507	52 500	3938
0.07	50 900	3856	31 500	2678
0.10	35 000	2917	0	0

possibility already at a discount rate of 10 per cent. This is due to the rather low growth rate implied by the parameter values used; the maximum growth rate is 10 per cent ($a = 0.1$). Most fish stocks are likely to have a much higher rate of growth at their most productive levels. Note that this example serves the purpose of underlining the powerful effect of the discount rate when all else is equal, and the importance of how sensitive the unit cost of fish is to changes in the fish stock. There are other considerations besides the discount rate that have to be made when it comes to irreversible decisions such as extinction of stocks, as discussed in Chapter 2.

Secondly, whether or not unit costs are stock-dependent is seen to have a very significant effect on the optimal stock size, particularly at high rates of discount. When the unit cost is inversely proportional to the size of the stock, the optimal stock size is well above zero even if the discount rate is as high as the maximum growth rate (10 per cent). This drives home again the point that a rising unit cost as a stock becomes depleted provides some protection against exploitation for fish and game stocks.

The consideration above of the impact of the discount rate is somewhat partial, as it does not deal with the impact of the discount rate on fishing costs. A higher discount rate means a higher cost of capital and hence increased fishing costs, particularly in capital intensive technologies (Hannesson 1987). This will to some extent counteract the effect that a higher discount rate causes the optimal stock to become smaller, as a higher cost per unit of effort (in the stock-dependent unit cost case) has the opposite effect.

3.3 The Ricker model

The catch from some fish stocks is based on one or a few mature year classes of fish on their way from the feeding grounds to the spawning areas. This is particularly true of salmonid species like salmon and capelin. A model that relates

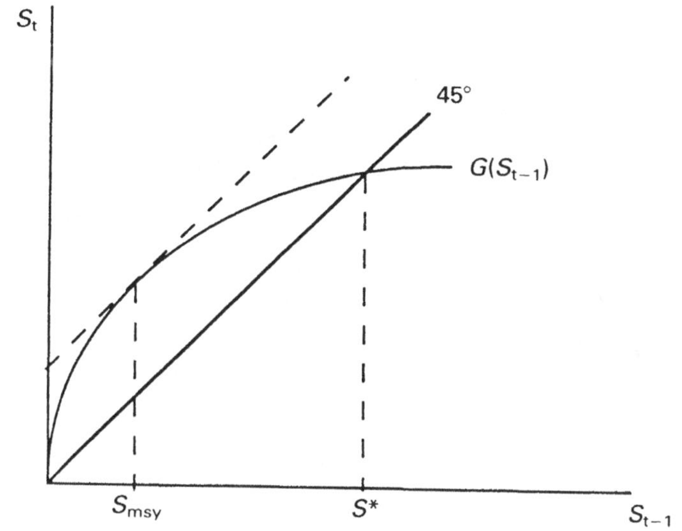

Fig. 3.2 A discrete growth function.

the size of the returning year classes (the spawning runs) to the size of the spawning stock a certain time period earlier is appropriate in this case. One such model was developed by the Canadian fisheries biologist W.E. Ricker (Ricker 1975). Since both spawning and fishing are seasonal, lasting only a few weeks each year, it is natural to use a discrete model.

A general expression for discrete stock-growth models is

$$S_t = G(S_{t-1}) \tag{3.14}$$

This relates the size of the stock in period t to the size of the stock one period earlier. The choice of a period is a matter of convenience, but is also influenced by the nature of the growth process. The appropriate length of the time period in stock–recruitment models is the time that lapses from spawning to the return of mature fish. This period usually is several years. The symbol G does not have the same meaning here as in the continuous model in the previous section and Chapter 2, where it meant surplus growth. The surplus growth in this case is $S_t - S_{t-1} = G(S_{t-1}) - S_t$

Figure 3.2 shows the growth curve $G(.)$. Clearly it must be above the 45° line for (at least some) stock levels below the natural equilibrium (S^*). The 45° line, sometimes called the replacement line, shows the stock level in period t needed to maintain the stock at the same level as in period $t - 1$. Any growth in excess of what is needed for replacement would increase the stock. The natural equilibrium is the point where the growth curve intersects the 45° line, and the maximum

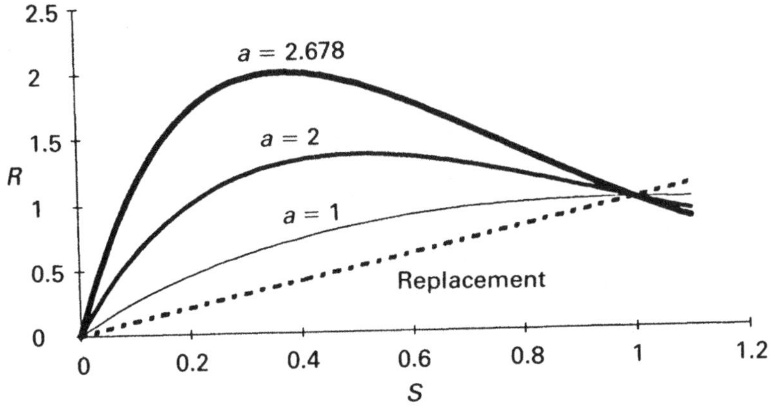

Fig. 3.3 Recruitment function. $R = S \exp[a(1 - S)]$.

surplus growth (growth in excess of replacement) is where the tangent to the growth curve has a slope equal to unity, parallel to the 45° line.

The slope of the tangent line to the growth curve is

$$dS_t/dS_{t-1} = dG(S_{t-1})/dS_{t-1} = G'(S_{t-1})$$

writing $G'(.)$ for the first derivative of the function $G(.)$. Clearly, if $G(.)$ is to lie above the 45° line, its slope must be positive and greater than unity for small S. As S increases the slope of $G(.)$ must gradually diminish, and at the natural equilibrium it must be less than unity. It could, however, be either positive or negative at the point of natural equilibrium, as shown by the various possible growth curves in Fig. 3.3.

Requiring that the natural equilibrium be stable imposes further restrictions on the slope of the growth curve. This is illustrated in Fig. 3.4, which is an enlargement of the area around the natural equilibrium. Suppose the growth curve is so steep at this point that its slope is greater than unity in absolute value (less than -1). Any small displacement of the stock from its natural equilibrium would send it further away from this point, as shown in the figure. Stability of natural equilibrium thus requires that the slope of the growth curve at the natural equilibrium point be less than unity in absolute value. The wisdom of imposing this requirement is, however, more doubtful than the requirement that a natural equilibrium exist. The equilibrium point could be unstable without the population having to become extinct as a result of a displacement from the natural equilibrium point. In Fig. 3.4 the deviations from the natural equilibrium become smaller as we get further away from the natural equilibrium, because the slope of the growth curve changes, and we end up with a cyclical movement around the equilibrium. It is not at all unlikely that animal populations undergo such movements, and some in fact seem to do so.

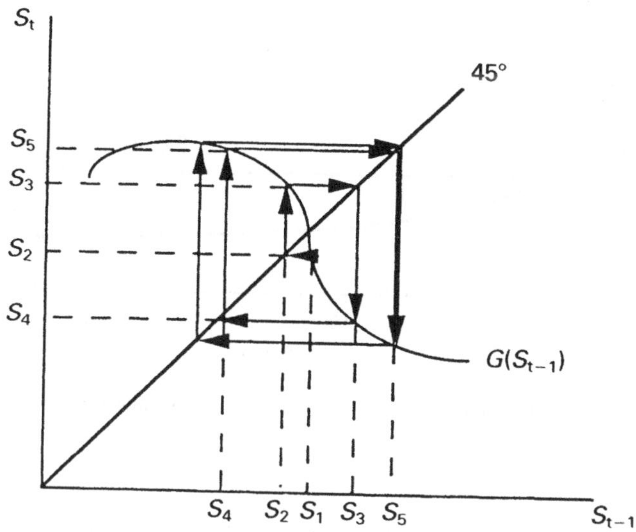

Fig. 3.4 Deviation from an unstable natural equilibrium in a stock–recruitment model. The deviation becomes larger and larger until a stable cycle obtains.

This discussion about the shape of the growth curve can be summarized as follows:

$G'(S^*) < 1$ (existence of natural equilibrium),
$G'(S^*) > -1$ (stability of natural equilibrium),
$G'(S) > 1$ for some $0 < S < S^*$,
$G'(S_{msy}) = 1$ (maximum surplus growth).

3.3.1 *Fishing in the Ricker model*

In the Ricker model, fishing and biological growth are treated as if they occurred sequentially and not simultaneously. This is an appropriate simplification when fishing is concentrated in a relatively short season during which there is little or no growth of individual fish and little natural mortality as well. To arrive at the stock available in period t we must subtract the catch taken in period $t - 1$ from the stock in period $t - 1$. Here we have two choices. We can either subtract the catch taken in period $t - 1$ from the stock at the beginning of period $t - 1$ (S_{t-1}), or the stock at the end of period $t - 1$, after we have allowed for the biological growth in period $t - 1$ [$G(S_{t-1})$]. In stock–recruitment models, where G is the number or biomass of fish emerging from eggs that were spawned one period earlier, it is appropriate to let G depend on the number (or weight) of fish that escaped fishing, provided that fishing takes place prior to spawning. We therefore subtract

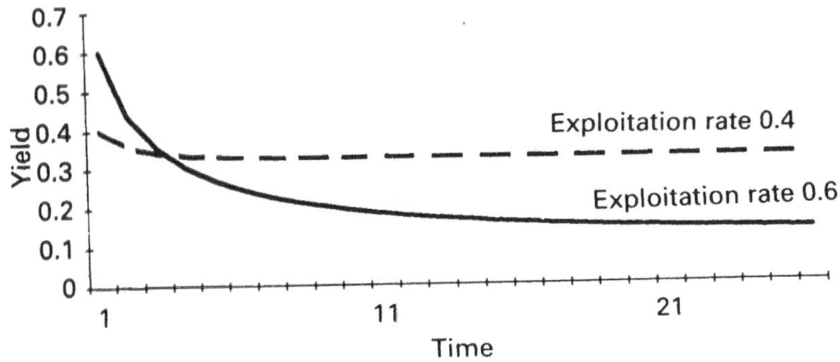

Fig. 3.5 Evolution of a fishery taking a constant share of the spawning stock in the Ricker model. $a = 1$, $S^* = 1$.

the catch in period $t - 1$ (Y_{t-1}) from the stock at the beginning of period $t - 1$, and Equation (3.14) becomes

$$S_t = G(S_{t-1} - Y_{t-1}) \qquad (3.14')$$

As in the continuous case, we may define sustainable yield as the surplus growth of the stock. Note that in the figures above, the variable S_{t-1} on the horizontal axis must now be replaced by $S_{t-1} - Y_{t-1}$, since it is the stock left after fishing that determines the size of the stock at the beginning of the next period.

When the relationship between the stock left behind in a given period and the stock one period later depends on the spawning stock–recruitment relationship, the function $G(.)$ is more appropriately called recruitment function. Ricker's proposed formulation of the recruitment function is

$$G(X) = X \exp[a(1 - X/S^*)] \qquad (3.15)$$

where $X = S - Y$ is the escapement (the size of the spawning stock). When there is no fishing and the spawning stock is identical to the natural equilibrium stock, we get $G(S^*) = S^*$ and the stock reproduces itself exactly.

The first derivative of the Ricker function is

$$G'(X) = (1 - aX/S^*)\exp[a(1 - X/S^*)] \qquad (3.16)$$

As explained above, the existence of natural equilibrium requires $G'(S^*) < 1$, which implies $a > 0$, while stability of natural equilibrium requires $G'(S^*) > -2$, which implies $a < 2$. Figure 3.3 shows various possible forms of the Ricker curve and how they depend on the value of a.

Figures 3.5–3.7 show the development through time of fisheries that start from a pristine stock and thereafter take a certain fraction of each 'spawning run', for alternative values of the parameter a (same as in Fig. 3.3). Figure 3.5, which is based on $a = 1$, shows large initial catches which fall over time and become

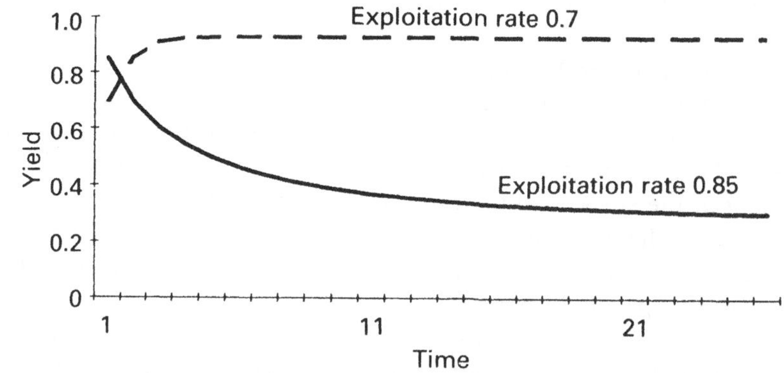

Fig. 3.6 Evolution of a fishery taking a constant share of the spawning stock in the Ricker model. $a = 2$, $S^* = 1$.

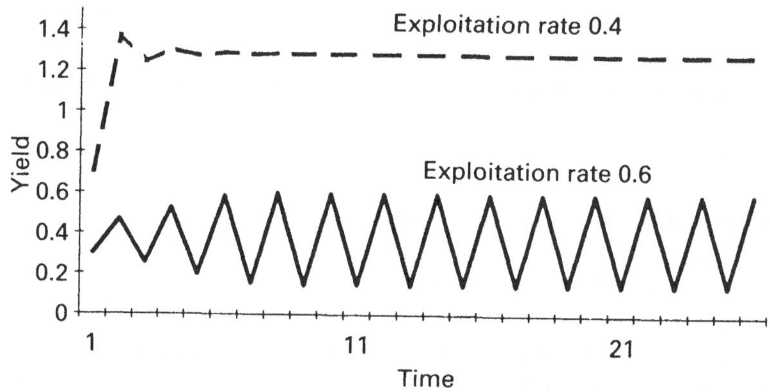

Fig. 3.7 Evolution of a fishery taking a constant share of the spawning stock in the Ricker model. $a = 2.678$, $S^* = 1$.

stabilized at a sustainable level. Figure 3.6 is based on a higher value of a ($a = 2$). Here it happens that for 'intermediate' rates of exploitation the catch increases from the beginning up to a sustainable level. The reason is that the growth curve has a high 'hump', implying that intermediate stock levels are very productive compared with the natural equilibrium level (the curve in the middle in Fig. 3.3). Figure 3.7 shows an unstable population with $a = 2.678$. Here the population can in fact be stabilized by selecting a rather high rate of exploitation, while low rates of exploitation give rise to a 'boom and bust' path of catches which develops into a stable cycle.

3.3.2 *Optimum stock in the Ricker model*

The relationship between fishing effort and catch will, as in the continuous model, depend on the distribution of fish in the sea and how it may change as a result of fishing. We shall make further use of the two stylized cases that we have used previously, the one in which the catch per unit of effort is always constant and independent of the stock, and the other where it is proportional to the size of the exploited stock. The case with a constant catch per unit of effort is essentially the same as before, and the shape of the yield curve will be similar. The other case is slightly more complicated, as we now have to take into account the continuous thinning of the population while fishing is going on (in the continuous case the fish stock is continuously replenished and so always stays the same in equilibrium). We shall deal here with the case where the cost per unit of effort is constant and, as earlier, denoted by c. Using the 'Schaefer' production function (3.3), the unit cost of fish caught at a given level of stock will be

$$cZ/ZqS = c/qS$$

The catch over a period is equal to the reduction in the stock over the period (ignoring natural mortality). Denoting the stock at the beginning of a period by S and the stock at the end of a period by X (the spawning stock), we get

$$Y = \int_x^S du = S - X$$

and the total cost of the catch over the period is

$$C(S,Y) = \int_x^S (c/qu)du = c(\ln S - \ln X)/q \qquad (3.17)$$

We can derive the conditions for the economically optimal level of the stock remaining at the end of each period (X) by considering a small deviation from that value. The deviation must be such that no gain can be obtained by departing from the said position; otherwise it could not be the economic optimum. Hence the short-term gain must be equal to the long-term loss. The short-term gain will occur in the period when the change in the stock level takes place. This gain will be equal to

$$-(p - c/qX^o)\Delta X$$

The term ΔX is the proposed change in the stock, and the change in the catch must be of the opposite sign, hence the minus sign. The price is constant and equals p, and the unit cost when the stock has been driven down to the optimal escapement level X^o is c/qX^o, so the cost of increasing the catch by $-\Delta X$ when that level has been attained is $-(c/qX^o)\Delta X$.

The present value of the infinite stream of optimal catches from the next period onwards is

$$\{p[G(X^o) - X^o] - c[\ln G(X^o) - \ln X^o]/q\}/r$$

The change in this as a result of changing X by ΔX is:

$$[p(G'(X^o) - 1]\Delta X - c\{[G'(X^o)/G(X^o) - 1/X^o]/q\}\Delta X/r$$

Setting the sum of the long- and short-term effects equal to zero and cancelling ΔX gives

$$(p - c/qX^o)(r + 1) - G'(X^o)[p - c/qG(X^o)] = 0$$

which we may write as

$$G'(X^o) = (r + 1)(p - c/qX^o)/[p - c/qG(X^o)] \tag{3.18}$$

With a constant unit cost of fish (c_y), the gain in the first period becomes $-(p - c_y)\Delta X$, while the change in the present value of sustained yield will be $(p - c_y)[G'(X^o) - 1]\Delta X/r$. Again setting the sum of the short- and long-term effects to zero gives

$$G'(X^o) = 1 + r \tag{3.18'}$$

From this we can derive the same qualitative results as we did in Chapter 2 for the continuous model. When the unit cost of fish is independent of the stock, the optimum stock level is where the slope of the tangent to the growth curve is equal to $1 + r$. This implies an optimum stock size that is less than the stock giving the maximum sustainable yield (S_{msy}, see Fig. 3.2). If the unit cost of fish is stock-dependent, the term $1 + r$ on the right hand side of (3.18) is multiplied by the term

$$(p - c/qX^o)/[p - c/qG(X^o)] < 1, \qquad (\text{since } G(X^o) > X^o)$$

which implies a greater optimum stock, possibly greater than that which gives the maximum sustainable yield.

Let us illustrate by a simple numerical example. Let $p = 1$, $a = 1$, and $S^* = 1$. Table 3.3 shows the optimum stock size for various values of the discount rate and the cost parameter c/q.

In the case shown in the table, the optimum stock size is not very sensitive to realistic variations in the discount rate, but the effect of a positive rate of

Table 3.3 Optimum stock in the Ricker model as a function of the discount rate (r) and cost of exploitation (c/q). $a = 1, p = 1$.

r	$c = 0$	$c/q = 0.2$	$c/q = 0.5$
0	0.430	0.510	0.649
0.05	0.414	0.498	0.642
0.10	0.398	0.486	0.637
0.15	0.381	0.475	0.632

discount is to push the optimum stock level below the maximum sustainable yield level ($X = 0.43$) when the unit cost of fish is constant (same as the costless case, $c = 0$). On the other hand, the effect of a stock-dependent unit cost of fish on the optimum stock level is very significant. The value of the parameter c/q shows at which stock level further depletion becomes uneconomical; this happens when the unit cost of fish becomes equal to the price, that is

$$c/qX = p$$

which implies $X = c/q$ for $p = 1$. Even at $c/q = 0.2$ the optimum stock size is well beyond the maximum sustainable level at all realistic rates of interest.

3.4 The Beverton–Holt model

A year class model for fish stocks was developed by the British biologists Raymond Beverton and Sidney Holt in the 1950s (Beverton & Holt 1957). This model is widely used for fish stocks in temperate waters, as these typically consist of many year classes of fish. The model has been extensively modified since it was first developed, mainly for making it better suited for computers, but the approach is essentially the same as that taken by Beverton and Holt.

The basic idea behind the model is to follow each year class of fish from its recruitment to the fishable stock to its virtual disappearance, explaining the yield from the stock by the gradual reduction of the year class through fishing, natural mortality, and the growth of individual fish. It is natural to start with the change in the number of fish, N, in a given year class. The rate of change at a point in time is

$$dN/dt = -(F + M)N_t \tag{3.19}$$

The expression dN/dt means the number of fish that are disappearing per unit of time. The value of this expression depends on how quickly the fish are disappearing from the stock and how large the stock is. If the stock is disappearing very rapidly, like stocks of farmed fish that may be harvested over a few hours or days, the value of dN/dt approaches infinity. Suppose, for example that there are 10 000 fish in a pond and that we scoop them up in 8 hours. The value of dN would be $-10\,000$ and the value of dt 0.0009132, measuring time in years, while dN/dt would attain the impressive value of 10 950 000. A high value of dN/dt thus means that the fish are removed from the stock very quickly, and if they are removed virtually instantaneously, the value of dN/dt approaches infinity.

The parameters F and M denote instantaneous fishing and natural mortality and determine how quickly the fish disappear from the stock. In the pond example just mentioned the value of F would be approximately 1095. This is not likely to be typical of ocean fisheries, but still the value of F can be quite high; values above 1 are not at all unrealistic and appear to occur for many heavily exploited stocks. Note that $F = 1$ does not imply that the stock will be wiped out in one year, as it would do if we were dealing with mortality per year and not instantaneous mortality. This can be seen by calculating e^{-1} (see Equation (3.20)). The value of M depends on predation by other fish and the physical condition of the fish (state of health and nutrition). At certain stages in the life of the fish this mortality can be very large, as at the larval stage and after spawning. Most fish of certain species die after spawning (Pacific salmon, for example), in which case M approaches infinity, like F in the pond example.

Finding the total number of fish that remain in the stock after a certain period of time such as one year is accomplished by solving the differential equation (3.19). Letting N_t denote the number of fish at the beginning of period t and N_{t+1} the number of fish at the beginning of period $t + 1$, we get

$$N_{t+1} = N_t \exp(-F - M) \tag{3.20}$$

The catch of fish in terms of numbers can be found similarly by considering the fish that disappear because of fishing. In total, $N_t - N_{t+1} = N_t[1 - \exp(-F - M)]$ have disappeared from the stock from time t to time $t + 1$. Of this the fraction $F/(F + M)$ must have disappeared because of fishing. Hence the total catch in numbers over the year is

$$[F/(F + M)]N_t[1 - \exp(-F - M)] \tag{3.21}$$

We can find the catch in terms of weight by multiplying this by the average weight each year of a representative fish in each age group, $w(h)$, where h denotes the age of the fish. This gives

$$[F/(F + M)]N_t[1 - \exp(-F - M)]w(h) \tag{3.22}$$

Proceeding in this way involves an error if the fish grow continuously over the year. The appropriate value of $w(h)$ to use depends on how rapidly the fish are being removed; if they disappear very rapidly, as in the pond example above, then clearly we should use the weight of the fish at the point in time when we empty the pond. The original formulation by Beverton and Holt incorporated a growth function of the fish, so that this problem did not arise. In ocean fisheries the inaccuracy arising from using the average weight of fish of a certain age is not likely to be very great. The inaccuracy that could arise would be due to a substantial change in fishing mortality. An increase in fishing mortality would cause a more rapid depletion of the stock, and the fish caught would therefore weigh less because they would have had less time to grow before being removed. If a greater precision is desired, this could be accomplished by a finer division of time, updating the model quarterly or monthly or whatever, but still taking the fixed $w(h)$ approach.

To find the total catch of fish from a stock we must sum the catch from all year classes in the stock. This gives us the Beverton–Holt yield function:

$$Y_t = \sum_{h^*}^{h^{**}} [s_h F/(s_h F + M)] N_{ht} [1 - \exp(-s_h F - M)] w_h \tag{3.23}$$

where h^* is the age at which the fish are recruited to the stock and h^{**} is the age at which they disappear. The choice of h^{**} is essentially a practical one; individual fish can apparently attain an exceptionally high age, but how soon a year class is virtually gone from the fishable stock depends on how high the mortality is. If we want to investigate the consequences of large variations in fishing mortality, it is desirable to determine h^{**} on the basis of how many fish would be left in the year class if it had suffered only natural mortality; otherwise the model could seriously underestimate the gains of increasing the survival rate of fish.

In Equation (3.23) we have sneaked in a parameter s_h. The purpose of this parameter is to modify the value of the fishing mortality parameter according to the age of the year class. Young fish typically are less exposed to fishing than older fish; they may slip through meshes, for example. Experiments with trawls with meshes of different size have shown that changes in mesh size affect how soon fish become liable to capture. Mesh size regulations therefore are among the important issues of fisheries management. Hence the value of s_h, which is called gear selectivity, will typically vary from zero (the fish are not captured at all) to 1 (the fish are fully vulnerable to the fishing gear). It is possible to have s_h reflect spatial distribution of fish as well; that is, the year classes are present on the fishing grounds to a varying degree according to their age.

The size of each year class in year t is the result of previous fishing and so can be calculated by a formula like the following

$$N_{ht} = N_{rt-h+h^*} \exp\left[\sum_{j=h^*}^{h-1} - (s_j F_{t-h+j} + M)\right] \tag{3.24}$$

where N_r is recruitment (the number of fish originally in the year class at age h^*), and the sum is defined as zero for $h = h^*$.

As an illustration, translate this formula into words:

Suppose the fish are recruited at the age of 3. The number of 7 year old fish at the beginning of 1991 then equals the recruitment in 1987, minus the number of fish that have died. The fish that have died were exposed to natural mortality for 4 years (1987–90), plus fishing mortality of 3 year old fish in 1987, 4 year old fish in 1988, 5 year old fish in 1989, and 6 year old fish in 1990.

3.4.1 *Sustainable yield*

We are now equipped to consider sustainable yield and optimal stock size, etc., in the Beverton–Holt model. The first thing to note about sustainable yield is that, for a sustained fishing mortality and no fluctuations in recruitment, the stock will be in an equilibrium. Hence its age structure must remain the same over time,

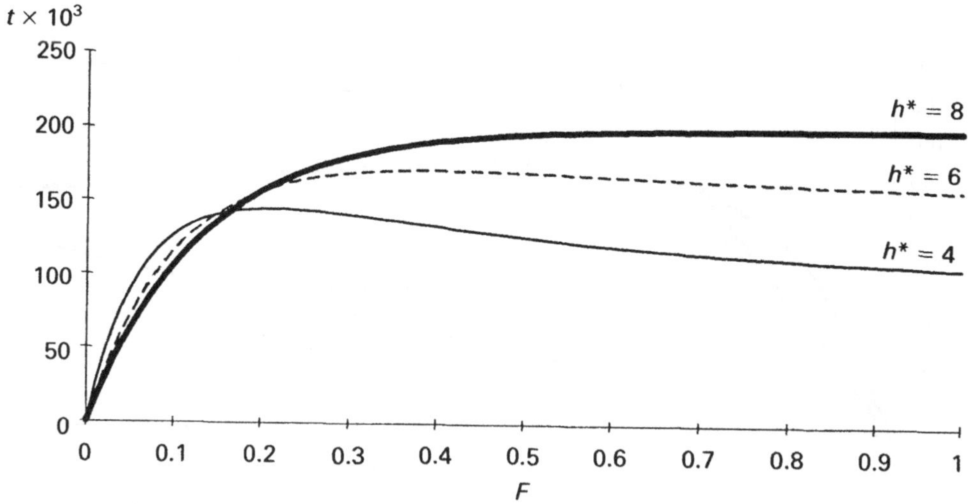

Fig. 3.8 Sustainable yield in the Beverton–Holt model. Data for Arcto-Norwegian cod. h* = age at first capture.

which implies that the catch of fish from the stock in any given year must be the same as the catch from a single year class over its lifetime. Hence we may use the formula

$$Y = \sum_{i=h^*}^{h^{**}} [s_iF/(s_iF + M)]\{1 - \exp(-s_iF - M)\} \exp\left[-\sum_{j=h^*}^{i-1} (s_jF + M)\right]N_r w_i \tag{3.25}$$

where F is the constant fishing mortality, N_r is the number of recruits, and the sum in the exponential term is defined as zero for $i = h^*$.

Figure 3.8 shows a family of sustainable yield curves for Arcto-Norwegian cod. The values of the parameter w (weight at age) were taken from one of the stock assessment reports of the International Council for the Exploration of the Sea (Arctic working group) and are as follows (weight in kg × 10^2):

h	4	5	6	7	8	9	10	11	12	13	14	15+
w	100	155	235	345	470	617	770	925	1085	1250	1390	1500

The curves shown in Fig. 3.8 were derived by varying the parameter s_i, the age at which the fish are selected by the fishing gear. Knife-edge selectivity was assumed, so s is either zero or one. The recruitment was set at 100 million individuals per year. This is at the lower end of the range over which the recruitment of Arcto-Norwegian cod has varied. The sustainable yield reported below, which has a maximum of about 200 000 t, does not reflect a realistic average yield but serves only an illustrative purpose.

What, then, are the optimum sustainable fishing mortality and fish yield?

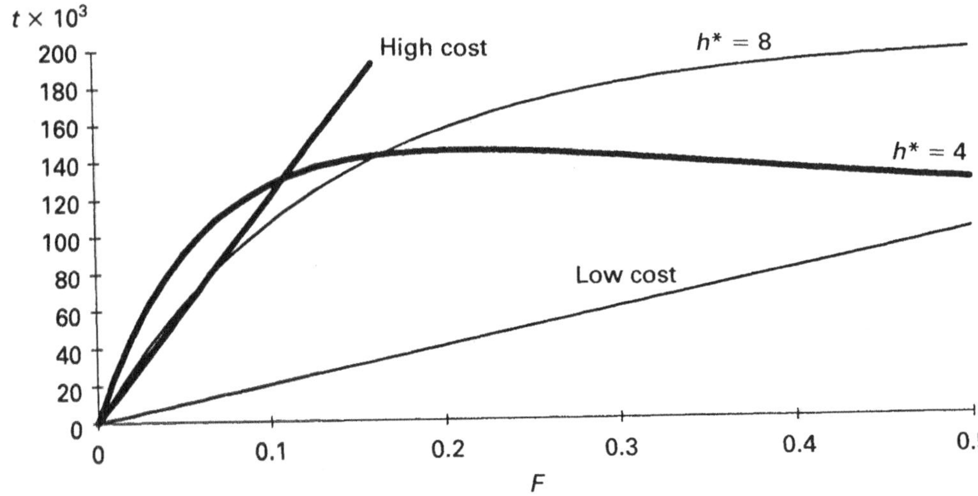

Fig. 3.9 Optimum sustainable yield. 'High' and 'low' cost of effort.

Figure 3.9 reproduces portions of two of the sustainable yield curves from Fig. 3.8, the one with 4 and the other with 8 as the age at which the fish are selected (age at first capture). For a constant price, set equal to one for convenience, these curves also show the value of the sustainable yield. The figure also shows two cost lines. The shape of these lines implies that the cost per unit of effort is constant and that fishing mortality (F) is proportional to fishing effort. Since fishing mortality is the proportion of the stock being removed by fishing during a 'short' period of time ($F = Y/S$), this in turn implies that the catch per unit of effort is proportional to the stock ($F = qZ$; $Y/Z = qS$), one of the stylized cases we have discussed extensively above. Maximizing the sustainable rent (the distance between the sustainable yield curve and the cost line) is seen to involve choosing one or the other of the yield curves, depending on how high the cost per unit of effort is. If the cost per unit of effort is high (the steep cost line), the optimum age of the fish at first capture is low, and the optimum fishing mortality is also low. If the cost per unit of effort is low (the lower cost line), the optimum age at first capture is high, and the optimum fishing mortality is also high.

Maximizing the sustainable rent thus involves two decisions, selecting the optimum level of F and the best age at first capture. Fishing mortality can in principle be controlled by controlling the activity of the fishing fleet, but how can we influence the age at first capture? As mentioned earlier, the selectivity of the fishery can be varied by changing the design of the fishing gear, such as meshes in nets and trawls. This cannot be done so neatly as to select only fish above a certain age; the year classes of fish typically are selected gradually with age, with the values of the parameter s increasing gradually from zero towards one. Even so, the principle is the same as in the case of knife-edge selectivity portrayed in Fig. 3.8.

It is possible to give an intuitive explanation of the interrelationship between optimum fishing mortality and optimum age at first capture, both of which decrease when the cost per unit of effort rises. As the fish become older, more and more fish are lost because of natural mortality. If we were concerned only with the number of fish we catch, we would catch them as quickly as possible at their youngest accessible age. But what is important is how many tonnes of fish we catch, not how many fish. The maximum aggregate weight of a year class is attained at the age when the relative growth rate of individual fish is equal to natural mortality. This can be seen by differentiating the aggregate year class biomass, $W(h) = N(h)w(h)$, with respect to age (h). This yields $dW/dh = w(dN/dh) + N(dw/dh)$. Setting this equal to zero gives $(dw/dh)/w = -(dN/dh)/N = M$.

Since fish grow with age but at a decreasing rate, it would be preferable to wait until they had reached the age of maximum year class biomass, provided that we could take the whole year class (or as much of it as we want to) instantaneously at that age. This would require a very high fishing mortality at that particular age. If the cost per unit of fishing mortality is high this would not be profitable. Applying a 'low' fishing mortality and selecting only year classes which have reached the age of maximum aggregate weight, we would lose a fair amount of the potential catch from each year class, because we would not have caught any sizeable portion of it until it had reached an age at which much of its biomass will have been lost through natural mortality. It would be better, therefore, to start selecting the fish at a younger age, to minimize the loss. Clearly the loss arising from selecting the fish too late will be less the greater is the fishing mortality. Hence, the less the cost per unit of effort and the higher the optimum fishing mortality, the higher is the optimum age at which the fish should be selected by the fishing gear.

If the cost of fishing is proportional to the quantity caught (constant catch per unit of effort), the total cost in Fig. 3.9 will be represented by a horizontal line (not drawn). The total cost would then be independent of the fishing mortality. This is the other of the two stylized cases we have used above. Maximizing the sustainable rent is in this case identical to maximizing the sustainable yield, which means attaining the highest of all possible yield curves and fishing the selected age groups with maximum intensity (F approaches infinity, as in the pond example mentioned at the beginning of this section). With knife-edge selectivity, this means selecting only the age group which has attained its maximum weight and fishing all that is available of it. When such fine tuning is not possible maximum sustainable rent implies attaining the highest possible yield curve and choosing the F-value corresponding to the top of it.

At this point it is natural to discuss some important aspects glossed over by the sustainable yield concept as we have presented it here. Firstly, it ignores any relationship there might be between the recruitment of young fish and the size of the parent stock, as recruitment was assumed constant. Even if this assumption is defensible for moderate variations in the size of the spawning stock, as there seems to be a weak correspondence between the variations in the size of the spawning stock and the recruitment of young fish, this cannot be defended if fishing intensity becomes so great that the year classes are almost or wholly wiped out before they reach the age of spawning. As we approach the tail end of the yield curves in Fig. 3.8, where fishing mortality becomes very high, this could occur, depending on whether or not the selectivity of the gear is such that it allows the fish to survive to spawning age before they become vulnerable to fishing.

Another shortcoming of the sustainable yield curves presented in Fig. 3.8 is that they ignore possible density-dependence of growth and mortality. Both weight at age and natural mortality are assumed to be the same whether a year class (or the total population) is large or small. Growth of individual fish is, however, quite likely to be density-dependent, as fewer fish in a year class would mean less competition for food, which is likely to make fish grow faster. As to mortality, cannibalism is a phenomenon that often occurs in fish stocks. If the fishing mortality is low, old year classes will be relatively plentiful, causing more predation than otherwise on the younger year classes. These growth and mortality effects imply that the yield curve will have lower values for low rates of fishing mortality (fish live longer on the average but weight less, and more of the younger fish are lost through predation) and higher values for higher rates of fishing mortality. In few if any cases, however, have such effects been quantified. At some 'high' rate of fishing mortality the yield could fall drastically, owing to the ravaging of cohorts before they reach spawning age. Correcting for this would imply a somewhat greater optimum fishing mortality than one would otherwise recommend, at least as long as the optimum is on the left side of the maximum of the sustainable yield curve. Discounting of future values may, as demonstrated earlier, imply that the optimum is on the right hand side of the maximum. We now turn to examining the consequences of discounting in this present model.

3.4.2 *Discounting in the Beverton–Holt model*

In Chapter 2 we tried to explain the impact of discounting of the future by looking at the short-term gains that can be accomplished by driving down the standing stock below the value that maximizes sustained annual rent and comparing these short-term gains with the present value of the loss of sustained rent. It is natural, therefore, to start investigating the impact of discounting in the Beverton–Holt model by considering a deviation from the value of fishing mortality that maximizes sustained rent.

Let us again use the model developed in Section 3.4.1. For illustrative purposes we shall use the selectivity pattern that takes fish 4 years and older ($s_i = 1$;

$i = 4, \ldots$). We shall look at the case where the costs of fishing are proportional to fishing mortality (F) and set the cost per unit of fishing mortality equal to 200, again for illustrative purposes. By using the model developed in Section 3.4.1, we can calculate the optimal value of F without discounting (the one that maximizes the vertical distance between the sustainable yield curve and a cost line with a slope of 200). This gives $F = 0.14$.

Now suppose the stock has been exploited at this rate for more years than the number of year classes of fish (of any significance) in the stock. We can then use Equation (3.24) to calculate the age composition of the stock (number of fish in each year class) at the beginning of next year, which we put as year 1. If we now increase the rate of exploitation from the next year onwards, we will get a higher rent for a number of years, but eventually we must end up with less rent per year than if we had kept the fishing mortality at $F = 0.14$. Calculating the F-value that maximizes the present value of future rents at the beginning of year 1 gives $F = 0.22$ for a discount rate of 10 per cent, which is greater than the F-value that maximizes annual sustainable rent (0.14). The present value of future rents is found by multiplying the rents each year by a discount factor $(1 + r)^{-t}$, where t is the 'serial number' of the future year starting with $t = 0$ in year 1, and summing the future rents. During the phase while the age composition of the stock is changing from the old equilibrium with $F = 0.14$ to the new one with $F = 0.22$ we must do this year for year, while after the stock has attained the new equilibrium age composition (which happens after it has been exploited at the new F-rate for as many years as there are year classes in the model) we can use the formula for a convergent geometric series to calculate the remainder. Denoting the annual rent by X_t and the present value of all future rents by PV we get

$$PV = X_1 + X_2(1 + r)^{-1} + \ldots + X_t(1 + r)^{-t+1} + \ldots X_k(1 + r)^{-k+1}/r$$

where k is the year when the stock attains the new equilibrium age composition.

Table 3.4 indicates why discounting of future rents has the effect of making an increased rate of exploitation more attractive. The table shows the annual rent starting with two different age compositions of the stock, a stock that has never been exploited and has suffered only natural mortality of $M = 0.2$, and a stock that has been exploited at a constant rate $F = 0.14$, as in the above example. We see that the annual rent is 'large' to begin with but drops off towards a long-term equilibrium. Maximum sustainable rent is 110.8, and we see that the long-term annual rent is less than that in both examples. During the first 7 years, however, the annual rent is greater. Because of the discounting the positive difference during the first 7 years cancels the present value of the negative annual differences occuring in later years.

If we start with a pristine stock, the age composition of which is determined by $M = 0.2$ and $F = 0$ in Equation (3.24), the optimum fishing mortality will be greater than in the other case, or $F = 0.33$. This is slightly more than the F-value that maximizes sustainable yield. This indicates that the optimum rate of

Table 3.4 Time profile of rents in a Beverton–Holt model where a constant fishing mortality has been set to maximize the present value of rents.

Rent per year	Initial stock determined by	
Year	$F = 0.14$	$F = 0$
1	166.2	677.1
2	151.4	475.4
3	139.2	337.0
4	129.3	243.4
5	121.5	181.1
6	115.5	140.4
7	111.1	114.3
8	107.8	97.8
9	105.4	87.6
10	107.8	81.4
.	..	..
.	..	..
∞	100.6	72.7

exploitation will depend on the initial size of the stock. But as the stock changes over time (because the fishing mortality has been changed), a new value of fishing mortality will become the optimum. This leads us to expect that the long-term optimum should be attained by adjusting the fishing mortality temporarily so as to attain the long-term optimum as quickly as is economically sound. Stocks that are too large must be fished down to the long-term optimum level. How quickly this should be done depends on costs. The more rapidly a stock is fished down, the more it is necessary to invest in fishing capacity which will not be needed when the long-term optimum has been attained. Hence, the higher the long-term cost of the fishing fleet, the more gradual the approach to the long-term optimum will be. The theoretical aspects of this problem are considered in Clark *et al.* (1979).

For overexploited stocks, the optimum strategy would be to stop fishing and resume it again when the stock has recovered to the optimum level. There are arguments for a gradual approach in this case as well. One such has already been mentioned, namely that the operating cost of existing boats is lower than the full cost of new boats. A lower cost implies a smaller optimum stock, which in turn implies that it would take less time to rebuild a depleted stock to the desired level. There are, however, additional reasons for not stopping the fishery completely, even if a simple bioeconomic model would tell us to do so. These reasons have to do with the maintenance of markets and of skilled labour, etc. There might be little point in waiting for a stock recovery if these essential elements for a profitable fishery disappear in the meantime.

To find the true long-term optimum rate of exploitation with discounting, we must find a value of F that gives us an age composition of the stock such that we would not want to change the value of F. By using the above formulas we can calculate this to $F = 0.2$. This gives an annual sustainable rent of 104.6, compared with the maximum sustained rent of 110.8 obtained by $F = 0.14$.

3.4.3 *Multigear fisheries*

It often happens that the age groups of fish are selected differently by different fishing fleets. This happens because different kinds of fishing gear have different selectivity patterns, or because different kinds of vessels fish in different areas. So-called 'inshore' and 'offshore' fisheries are an example of this. Some fish species migrate inshore in search of food or for spawning and can then be captured by small inshore boats. Bottom trawls typically fish rather indiscriminately, taking both young fish and older fish (this can, however, to some extent be controlled by changing the size of the meshes in the trawl). This may give rise to fisheries that are 'selective' and 'unselective', unselective fisheries capturing both old and young fish while selective fisheries are based on mature year classes close to maximum biomass age. The Arcto-Norwegian cod fishery is a good example of this. The trawl fishery takes place on the feeding grounds for younger fish and is based mainly on immature year classes. The coastal fishery is in part based on the spawning migration of year classes that have attained the age of maximum biomass.

The Beverton–Holt model is well suited to analysing this type of situation. By adjusting the parameter s_i it is possible to reproduce the selectivity pattern of each particular fishery. The yield in each fishery is obtained by a slight modification of the formula for the single fishery (Equation (3.23)). Suppose there are n fisheries, each with its own selectivity pattern. The yield of fishery j in time period t from a given stock will be

$$Y_{jt} = \sum_{h^*}^{h^{**}} \left[s_{jh} F_j \Big/ \left(\sum_{i=1}^{n} s_{ih} F_i + M \right) \right] N_{ht} \left\{ 1 - \exp\left(-\sum_{i=1}^{n} s_{ih} F_i - M \right) \right\} w_h \qquad (3.23')$$

The consequences for one particular fishery of an increase in effort in another fishery are exactly the same as from increasing the natural mortality of fish; there will be less fish available in the long term. How serious this interaction is depends on the selectivity pattern of the fisheries in question. If the effort is increased in an unselective fishery taking both old and young fish, the yield in the selective fishery taking only old fish will be strongly affected, because fewer fish will survive until the age at which they are taken by the selective fishery. Increasing the effort in a selective fishery will affect the unselective fishery much more lightly, because this will only increase the mortality of older fish, which are only a part of the stock on which the unselective fishery is based.

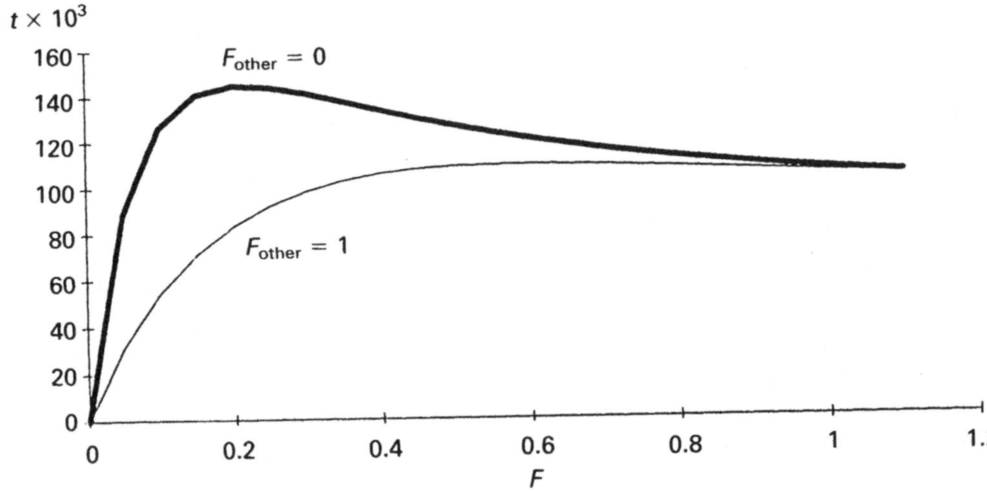

Fig. 3.10 Sustainable yield. Unselective fishery.

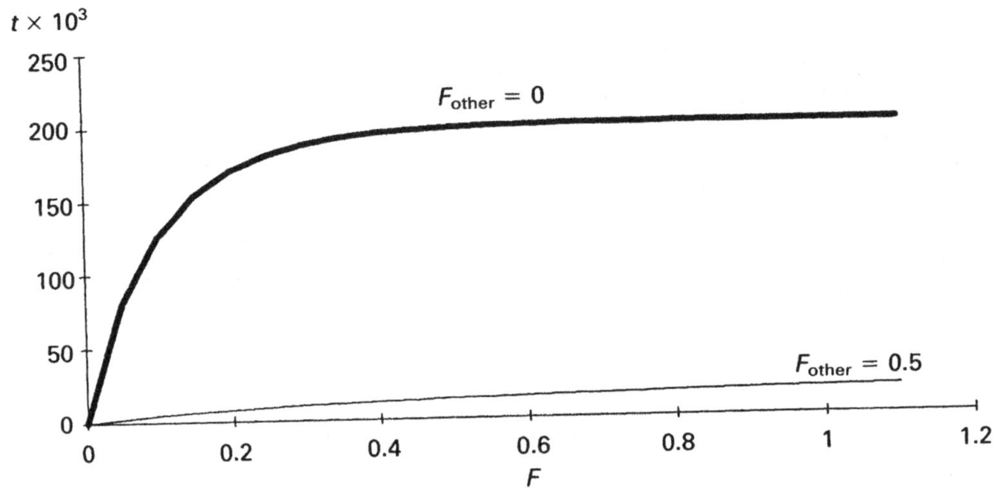

Fig. 3.11 Sustainable yield. Selective fishery.

Figures 3.10 and 3.11 illustrate this, using the same weight at age data as in the examples above. We assume only two fisheries, an unselective fishery taking fish 4 years and older indiscriminately, and a selective fishery taking only fish 8 years and older. Figure 3.10 shows two sustainable yield curves for the unselective fishery, one when there is only a single fishery (the unselective fishery) and another when the selective fishery produces a fishing mortality of 1.0. The selective fishery cuts the top off the sustainable yield curve of the unselective fishery and thus reduces its yield potential by almost one-third.

Both yield curves in Fig. 3.10 have the same asymptotic value. This means that a sufficiently great effort in the unselective fishery would eliminate the selective fishery altogether. The unselective fishery takes fish that are younger than those taken by the selective fishery. Hence, by increasing the fishing mortality produced by the unselective fishery, the year classes can be virtually eliminated before they become accessible for the selective fishery. The consequences of the unselective fishery for the selective fishery are illustrated in Fig. 3.11. The higher yield curve shows the sustainable yield of the selective fishery when there is no other fishery competing with it. The curve does not have a very pronounced maximum but has a greater asymptotic value than the maximum sustainable yield in the unselective fishery (see Fig. 3.10). This is so because the selective fishery takes only age groups that have attained their maximum biomass, while the unselective fishery also takes year classes that are still growing.

The lower yield curve shows the sustainable yield of the selective fishery when the unselective fishery produces a fishing mortality of 0.5. This reduces the yield of the selective fishery dramatically, by about 90 per cent. This does not mean that the unselective fishery is necessarily less profitable than the selective fishery; that depends on the cost structure, as discussed above, and prices. It could be the case, for example, that the unselective fishery is capable of producing better quality fish, or of sustaining more regular supplies, than the selective fishery. Still, the dramatic impact of the unselective fishery on the selective fishery indicates that fisheries managers could make serious mistakes if such interrelations among fisheries are not taken into account.

Chapter 4
Fluctuations in Fish Stocks

4.1 Introductory remarks

Fluctuations in the abundance of fish stocks are a well-known phenomenon. These fluctuations typically are associated with similar fluctuations in the catch that can be taken, or is allowed to be taken, from the stocks. This gives rise to two types of economic problems. First, fluctuating catches usually are inconvenient for the fishing industry. Existing markets may not be able to absorb increases in the quantity caught except at appreciably lower prices, while markets may be lost if deliveries cannot be made over a certain period. The processing industry, therefore, usually wants as stable catches as possible.

Second, the ability to take and to process large catches that occur only occasionally requires production capacity that will be partly idle most of the time. The economic life of durable equipment such as fishing boats and fish processing plants typically is much longer than the cycles of abundance of fish stocks. It makes limited sense, therefore, to invest in fishing fleets or processing plants that can accommodate large but occasional catches. If the catch cannot be stabilized, the question arises how large a catch it would be worth while to take.

In this chapter we consider these two problems. We begin by investigating whether in fact it would be desirable to stabilize the catch, even if it were possible. This is shown to depend on how the price of fish varies with the quantity caught and whether or not the catch per unit of effort depends on the size of the exploited stock.

Whether or not the catch can be stabilized and at what level depends critically on the possibility of 'saving' some of the fish available in 'good' years for later years with less fish. When a fishery is based on only one or a few year classes this is not likely to be desirable, and it may even be impossible. The catch would have to be stabilized at a rather low level, because the yield potential of the smallest year classes would have to be taken as a norm. For stocks consisting of many overlapping year classes of variable initial size that 'buffer' each other, a stable catch would be less dependent on the smallest year classes.

In the latter part of this chapter we consider how the optimal capacity of fishing fleets exploiting stocks with fluctuating catches depends on economic parameters and the variability of catches. First we consider this in a model with a randomly fluctuating stock and total allowable catch. The probability distribution of the stock and the total allowable catch is assumed to be given and stable over time. This implies that the amount caught in any particular year has no impact on the total allowable catch in later years. This is a useful simplification for bringing forth some basic principles for optimum investment in fleet capacity for fluctuating stocks. There are, however, stocks for which this is a reasonably accurate description of reality. Some short-lived species with a high turnover of biomass are accessible for fishing only after spawning has taken place, so that the amount removed from the stock has no influence on the recruitment to the stock, even if the number of recruits depends on the size of the spawning stock. The fact that the number of recruits also depends on environmental factors will then cause fluctuations in the size of the fishable stock which are in no way affected by the fishery.

Finally we consider the case where the future exploitable stock depends on how large a spawning stock is left behind, in addition to a random element reflecting environmental conditions. For this purpose we use a stochastic version of the Ricker model discussed in the previous chapter, but the principles derived apply equally to any other dynamic fishery model, such as the Beverton–Holt model for example.

4.2 Stable or variable catches?

Stabilization of the annual catch from a fluctuating stock is a feasible strategy only if the catch can be maintained constant without endangering the future recruitment to the stock. There is some presumption that stabilizing the catch would be risky in this respect. Maintaining a constant catch from a fluctuating stock implies that a greater proportion would have to be caught from a small stock than from a large stock. This could cause depletion of the spawning stock when the stock is small, which in turn would adversely affect recruitment. This strategy is particularly likely to be dangerous when the catch is based on one or a few year classes that vary greatly in size.

To demonstrate the possible economic advantage of stabilizing the catch from a variable stock, we shall assume that the amount of fish available in any given year is independent of the amount caught in previous years. This makes it easy to show why a stable catch might be desirable, but is a simplification of reality. As already mentioned, stabilization of the annual catch is most relevant for stocks the fishable part of which consists of several year classes. For such stocks, the quantity of fish available each year clearly depends on how much was caught in earlier years. Below we use the Beverton–Holt year class model to demonstrate numerically the possible benefit of stabilizing the catch. In this model the catch potential each year depends on previous catches.

In the preceding chapters we have relied extensively on two special cases of the relationship between catch and fishing effort, the first a constant catch per unit of effort and the second proportionality between the size of the fish stock and the catch per unit of effort. We shall continue to do so here. To repeat, the catch rate (catch per unit of time) in these two cases is

$$Y = k_1 Z \tag{4.1a}$$

$$Y = k_2 Z S \tag{4.1b}$$

where Y is the catch rate, Z is fishing effort, S is the size of the stock, and k_1 and k_2 are fixed coefficients.

We investigate the possible benefits of stabilizing the catch by comparing two catch strategies, a constant fishing mortality strategy and a constant catch strategy. Keeping the fishing mortality constant implies that a constant fraction of the stock is harvested. In the constant catch strategy the catch rate is set equal to the expected (average) catch rate under the constant fishing mortality strategy.

4.2.1 *The case of constant catch per unit of effort*

Consider first the case when the catch per unit of effort is constant (Equation (4.1a)). The constant fishing mortality strategy involves catching a constant fraction, k_3, of the stock:

$$Y = k_3 S \tag{4.2}$$

Since (4.1a) must be satisfied, this implies

$$Z = (k_3/k_1)S \tag{4.3}$$

The expected (average) catch will be (from 4.2)

$$EY = k_3 ES \tag{4.4}$$

where E denotes expected value.

By assumption, the constant catch strategy implies

$$\bar{Y} = EY = k_3 ES \tag{4.5}$$

which by Equation (4.1a) implies a constant effort:

$$Z = \bar{Y}/k_1 \tag{4.6}$$

Compare now the expected profit (π) from the two strategies. The profit per unit of time is the difference between the revenue per unit of time (R), which depends on the catch rate, and the cost per unit of time (C), which depends on the effort per unit of time:

$$\pi(Y,Z) = R(Y) - C(Z)$$

Using the above definitions, we get the expected profit under the two strategies:

Constant fishing mortality (Equations (4.2) and (4.3))
$E\pi(Y,Z) = ER(k_3S) - EC[(k_3/k_1)S]$

Constant catch rate (Equations (4.5) and (4.6))
$\pi(\bar{Y},\bar{Z}) = R(\bar{Y}) - C(\bar{Y}/k_1) = R(k_3ES) - C[(k_3/k_1)ES]$

In this special case, it turns out that both the revenue and the cost per unit of time will be constant when the catch rate is constant. The constant catch rate is by definition equal to a fixed proportion of the expected stock size (*ES*), and the effort is some multiple of the catch rate because of the assumed production relationship (Equation (4.1a)). The profit rate will therefore be constant. If, on the other hand, a constant fraction of the available stock is harvested, the catch rate will be uncertain because it fluctuates with the stock, and the effort per unit of time will likewise be uncertain because it fluctuates in proportion to the catch rate.

To see the economic implications of this, we must consider the cost and revenue functions. Note that the catch rate and the effort per unit of time are both multiples of the stock, so we can show the revenue, cost and profit in the same diagram as functions of the stock. This is done in Fig. 4.1. The shape of the functions in Fig. 4.1 corresponds to plausible assumptions about the cost and revenue relationships. The revenue is shown as a concave function of the catch rate (or the stock). This occurs if the price per unit of fish caught falls as the landed volume increases, but less in relative terms than the catch increases (this will be shown formally below). If the price is independent of the volume landed, the revenue function would be a straight line. The convex shape of the cost of effort function implies that the cost per unit of effort rises as more effort is applied, but in case the cost per unit of effort is constant the cost function would be linear.

Now suppose that the stock can assume only two values with equal probability, S_1 and S_2. The expected value of the stock, *ES*, will be half way between S_1 and S_2 (if S_1 were more likely to occur, *ES* would move closer to S_1). The profit of a constant catch rate equal to the fraction k_3 of the expected stock is shown by the point on the profit curve vertically above *ES*. The expected profit of the variable catch rate will be half way between the profit levels yielded by the two possible stock levels (S_1 and S_2), which is on the straight line connecting these two profit levels vertically above *ES*.

Clearly, as long as the revenue and cost curves are shaped the way they are in Fig. 4.1, the certain profit of the constant catch rate strategy will be greater than the expected profit of the constant fishing mortality strategy ($\pi(ES) > E\pi(S)$). If both of these relationships are linear, the two strategies will yield the same profit in the long run.

The shape of the revenue function is related to how the price of fish varies with the volume landed. To see how, consider the revenue function

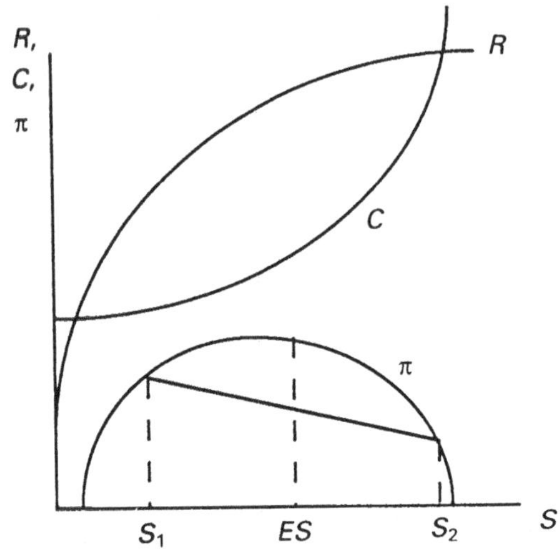

Fig. 4.1 Revenue (R), cost (C) and profit (π) as functions of stock (S).

$$R(Y) = p(Y)Y,$$

where p is the market price, which depends on the catch rate (volume landed). As p and Y move in opposite directions unless the price is independent of the quantity sold, R can either increase or decrease as the landings decrease. The price of fish and the volume of landings could increase together because of some additional factors simultaneously affecting both, such as a general increase in food prices and plentiful stocks of fish. Here we are looking at the relationship between price and quantity traded within a short period of time, other things being equal.

If the price rises relatively more than the volume of landings decreases, the revenue will rise as a result of a fall in the volume of landings. This can be shown mathematically as follows. Differentiating $R(Y)$ we find

$$dR/dY = p + Y(dp/dY)$$

$$d^2R/dY^2 = 2(dp/dY) + Y(d^2p/dY^2)$$

The shape of the revenue function shown in Fig. 4.1 implies that $dR/dY > 0$ and $d^2R/dY^2 < 0$. $dR/dY > 0$ implies $-(dp/dY)(Y/p) < 1$, which means that the relative change in price must be less in absolute terms than the relative change in quantity.

It is by no means impossible that the price will change more, in relative terms, than the quantity. The total revenue then increases when the quantity sold decreases and vice versa. This occurs in markets where consumers are willing to buy roughly the same quantity at quite different prices. Fish products are by no means unlikely to be in this category. If this is the case, the argument for stabilizing catches is reversed, unless the cost function is sufficiently convex.

Above we have compared the two fishing strategies in terms of average profit. In Chapter 2 it was argued that the sum of producer and consumer surplus is usually a more appropriate measure than profit of the social benefit. Here it may be said that we are taking a narrow industry view. Note, however, that we can use the same approach as here to evaluate the two strategies in terms of consumer and producer surplus; all we have to do is substitute consumer surplus (U) for revenue (R). The conclusions based on the concavity of the revenue function are particularly relevant in this context, because the consumer surplus is a concave function of the quantity sold (Y). This can be seen as follows:

$$U(Y) = \int_0^Y p(x)dx$$

$$\partial U/\partial Y = p(Y) > 0$$

$$\partial^2 U/\partial Y^2 = \partial p/\partial Y < 0$$

which implies concavity of U

4.2.2 *Catch per unit of effort proportional to the stock level*

In this case the production relationship is given by Equation (4.1b). As before, the constant fishing mortality involves catching a constant fraction, k_3, of the stock (Equation (4.2)), but now, by Equation (4.1b), this implies a constant effort:

$$\bar{Z} = k_3/k_2 \tag{4.7}$$

The expected catch under this strategy will be

$$EY = k_2\bar{Z}ES \tag{4.8}$$

Stabilizing the catch at $\bar{Y} = EY$ will, by Equation (4.1b), make it necessary to vary the effort inversely with the stock:

$$Z = \bar{Y}/k_2S$$

By Equation (4.8), the expected effort under this strategy will be

$$EZ = \bar{Z}ESES^{-1} \tag{4.9}$$

The profit of the two strategies now is:

Constant fishing mortality:

$$E\pi(Y,\bar{Z}) = ER(Y) - C(\bar{Z})$$

Constant catch rate:

$$E\pi(\bar{Y},Z) = R(\bar{Y}) - EC(Z)$$

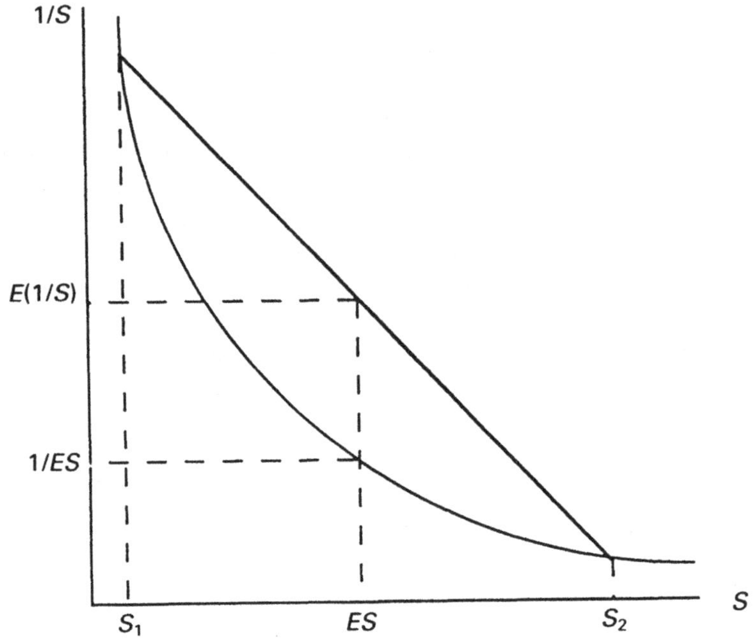

Fig. 4.2 The function $1/S$.

If the revenue function is concave (as in Fig. 4.1), $R(\bar{Y}) > ER(Y)$, which argues in favour of a constant catch rate. The cost relationship, however, now pulls in the opposite direction. Suppose, first, that the cost per unit of effort (c) is constant. Then,

$$C(\bar{Z}) = c\bar{Z} \quad \text{and} \quad EC(Z) = cEZ$$

By Equation (4.9), $EZ = \bar{Z}ESES^{-1}$. Now, S^{-1} is a convex function of S, as shown in Fig. 4.2, so $ES^{-1} > 1/ES$. Therefore, $ESES^{-1} > 1$, which implies $EZ > \bar{Z}$. More effort will thus be required on the average to take a constant catch than will be required to get the same catch on the average if the catch is allowed to fluctuate with the stock. This makes the constant catch strategy more costly for a given cost per unit of effort. If the cost per unit of effort varies progressively with effort, the constant catch strategy becomes more costly still, as the average cost of a variable effort will then be greater than the cost of a constant effort equal to the average effort (see Fig. 4.1). Hence we have

$$C(\bar{Z}) < EC(Z)$$

which argues in favour of the constant fishing mortality strategy.

Neither strategy thus is unambiguously better on strictly economic grounds. Which of the two is to be preferred depends on the parameters determining the

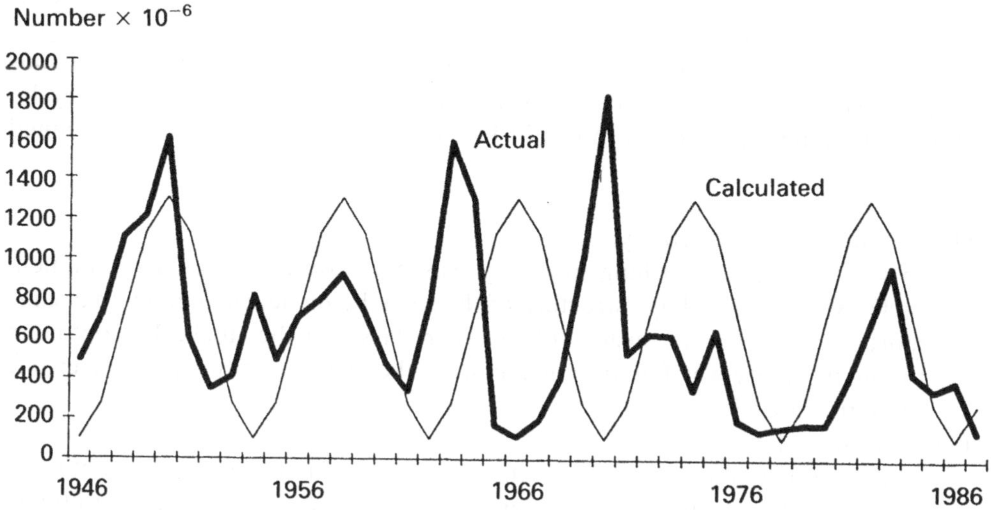

Fig. 4.3 Arctic cod recruitment.

shape of the cost and revenue function (or the consumer and producer surplus functions).

4.2.3 *An example of the Arcto-Norwegian cod*

The Arcto-Norwegian cod exhibits considerable variations in abundance which are due to variations in recruitment. These variations appear to be related to variations in sea temperature, and attempts have been made to relate them to changes in the earth's rotation axis (Larrañeta & Vazquez 1982). In addition, the growth of fish in the stock varies from year to year, presumably owing to variations in the supply of food.

In this exposition we will focus on the variations in recruitment and the resulting variations in stock abundance, ignoring the variations in fish growth. Fig. 4.3 shows the actual variations in the strength of year classes together with variations generated by an 8-year regular cycle. There is some regularity in the recruitment cycles; they appear to be of approximately 8-years duration, but there are obviously other factors affecting the recruitment pattern. The recruitment cycles are analysed in Hannesson and Steinshamn (1991), and the following exposition draws heavily on that paper. The cycles in Fig. 4.3 were generated by the sine function $N_t = N_{min} + u(1 + \sin v\pi t)$, where π has the usual geometric meaning. N_{min} is the minimum recruitment (100 million individuals). With u = 600 the maximum recruitment is $N_{max} = N_{min} + 2u = 1300$. With $v = 0.25$ the length of the cycle is $T = 2/v = 8$.

To investigate the economic implications of the two strategies discussed above, constant fishing mortality and constant catch, simulations of the Arcto-Norwegian cod stock were run, using the Beverton–Holt model discussed in the previous chapter. The recruitment process was assumed to follow a regular 8-year cycle, with a recruitment pattern as shown in Fig. 4.3. The two strategies were defined as follows:

(1) Constant fishing mortality:
In this strategy the fishing mortality was held constant over a recruitment cycle, with the initial age structure of the stock being determined by that same mortality over the life history of the year classes in the stock. The optimum fishing mortality was found by maximizing the average annual profit over one recruitment cycle.
(2) Constant catch:
In this strategy the catch is the same every year. The optimum constant catch was found by maximizing the average profit per year over one recruitment cycle. Keeping the catch constant makes it necessary to vary the fishing mortality inversely with the stock, fishing the stock most intensively when it is at its lowest level. Simulating the fishery over a long period generates a regular cycle of fishing mortality, and the size of the stock varies in a regular cycle. The initial stock was assumed to have been exploited in an optimum fashion for a long enough period to have settled down to a regular cycle of abundance.

The fact that we are looking at annual long-term (sustainable) profit implies that we are not discounting future profits, the rationale for which was discussed in Chapter 2. As discussed in Chapter 3, discounting makes the initial size of the stock and the approach to a long-term equilibrium important. These are not issues we want to discuss here; we are concerned instead with a comparison between two different fishing strategies. Some trial runs indicated that the discounting procedure is not important for that comparison. Looking at annual sustainable profit is the simplest way of dealing with the issue.

There are some noteworthy differences between the approach here and the theoretical model of the previous section. First, the fluctuations in the stock size are deterministic and correlated over time, instead of being random with a time invariant probability distribution like in the previous section. The theoretical results of the previous section are nevertheless relevant in this context. The frequency distribution of the stock size over one recruitment cycle may be substituted for the probability distribution in the previous section, and the conclusions of that section remain valid. Another difference is that we do not constrain the annual catch under the constant catch strategy to be equal to the average annual catch under the constant fishing mortality strategy. In most cases, however, the two are practically identical.

Table 4.1 Constant price of fish ($p = 1$) and cost per unit of fishing mortality ($c = 5000$).

h^*	4	5	6	7	8	9
Constant fishing mortality						
F	0.075	0.082	0.088	0.093	0.096	0.096
π	413.1	426.9	430.8	419.7	389.0	341.3
Y_{av}	788.1	836.9	870.8	884.7	869.0	821.3
Y_{max}	838.1	912.6	971.0	1004.7	1005.1	968.8
Y_{min}	738.1	761.2	770.7	764.8	732.9	673.8
X_{min}	7290	7420	7710	8161	8818	9583
Constant annual catch						
Q	790.0	840.0	870.0	880.0	860.0	810.0
π	413.3	426.6	430.0	417.0	384.2	334.5
F_{av}	0.075	0.083	0.088	0.093	0.095	0.095
F_{max}	0.080	0.090	0.099	0.107	0.112	0.114
F_{min}	0.070	0.075	0.078	0.081	0.081	0.079
X_{min}	7262	7330	7635	8142	8946	9841

h^* = age at first capture (knife-edge selectivity); F = fishing mortality; π = profit per year; Y_{av} = average catch per year; Y_{max} = maximum catch per year; Y_{min} = minimum catch per year; X_{min} = minimum size of the spawning stock; Q = annual constant catch; F_{av} = average fishing mortality per year; F_{max} = maximum fishing mortality per year; F_{min} = minimum fishing mortality per year.

Some results of the simulations are summarized below. We begin by comparing the two strategies when the catch per unit of effort is proportional to the stock size, and the price of fish and the cost per unit of fishing effort are both constant. The two strategies then yielded the results shown in Table 4.1. The theory in the previous section predicts that the constant fishing mortality strategy will be better; both will yield the same average revenue (because of the linearity of the revenue function), while the constant catch strategy will require greater effort on the average. This does not happen here, because we have not restricted the annual constant catch to be equal to the average annual catch under the constant fishing mortality strategy. What happens instead is that the average optimum effort in the constant catch strategy is less than the (constant) effort in the constant fishing mortality strategy, resulting in a (constant) catch that is smaller than the average annual catch in the constant fishing mortality strategy. The average annual profit of the constant catch strategy is less (except in one case) than that of the constant fishing mortality strategy, as concluded in Section 4.2.2, but the difference is very small, only about two per cent at the most.

Clearly, if a more significant difference between the two strategies is to be found, we must change some of the assumptions. The theory in Section 4.2.2 predicted that a concave revenue function would make the constant catch strategy

Table 4.2 Constant cost per unit of fishing mortality ($c = 5000$). Kinked revenue curve ($R = Y$ for $Y < Q$, $R = Q$ for $Y > Q$).

h^*	4	5	6	7	8	9
Constant fishing mortality						
F	0.067	0.072	0.074	0.075	0.076	0.074
π	410.0	421.8	422.5	408.7	376.4	329.0
Y_{av}	745.0	782.9	793.5	783.9	757.5	699.1
Y_{max}	789.5	849.6	878.2	881.2	864.4	811.3
Y_{min}	700.5	716.3	708.8	686.5	712.7	586.9
X_{min}	7855	8056	8532	9130	9790	10 539
Constant annual catch						
Q	790.0	840.0	870.0	880.0	860.0	810.0
π	413.3	426.6	430.0	417.0	384.2	334.5
F_{av}	0.075	0.083	0.088	0.093	0.095	0.095
F_{max}	0.080	0.090	0.099	0.107	0.112	0.114
F_{min}	0.070	0.075	0.078	0.081	0.081	0.079
X_{min}	7262	7330	7635	8142	8946	9841

h^* = age at first capture (knife-edge selectivity); F = fishing mortality; π = profit per year; Y_{av} = average catch per year; Y_{max} = maximum catch per year; Y_{min} = minimum catch per year; X_{min} = minimum size of the spawning stock; Q = annual constant catch; F_{av} = average fishing mortality per year; F_{max} = maximum fishing mortality per year; F_{min} = minimum fishing mortality per year.

more profitable. Table 4.2 shows the results of introducing a kinked revenue function (see Fig. 4.4), such that all catches in excess of the optimum constant catch are worthless. This is a rather extreme assumption, but serves the purpose of illustrating how large the difference between the two strategies could possibly be. The constant catch strategy now becomes more profitable, which accords with the theory in Section 4.2.2, but the difference is still small, or less than 2 per cent. Note that the difference between the two strategies would have been somewhat greater if we had stuck with the assumption made in Section 4.2.2, namely that the average catch under the constant fishing mortality strategy should be equal to the constant catch under the other strategy. Because all catches in excess of the optimum constant catch are worthless, it turns out to be a good idea to use less effort in the constant fishing mortality strategy than that which would yield an average annual catch equal to the optimum constant catch.

A noteworthy feature of the simulations summarized in Tables 4.1 ad 4.2 is that the minimum spawning stock is not much smaller in the constant catch strategy than in the constant fishing mortality strategy. This happens because the fishing mortality is very low in these simulations. With a higher optimal annual catch, the consequences for the spawning stock of keeping the catch constant could be disastrous. One such case will emerge below.

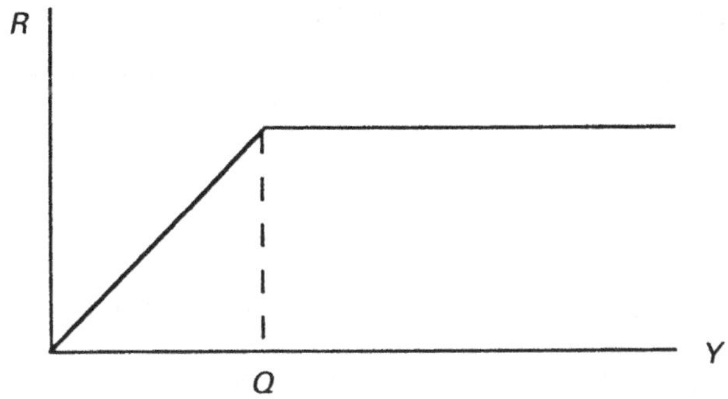

Fig. 4.4 A kinked revenue curve. *R* is revenue, *Y* is catch, and *Q* is an upper limit beyond which additional catch is worthless.

As discussed in Sections 4.2.1 and 4.2.2 above, the difference between the two strategies will be greatest and most favourable for the constant catch strategy if the catch per unit of effort is constant. We can deal with this case in a simple way by maximizing the value of the catch, as the cost of fishing will be proportional to the quantity caught and can be accounted for directly in the price of fish. Table 4.3 shows the result of using the same kinked revenue curve as shown in Fig. 4.4. The difference in the profitability of the two strategies is now up to 20 per cent. The spawning stock is wiped out in one case with a constant annual catch, which underlines the riskiness of the constant catch strategy. This occurs because the catch is fixed at a rather high level, so in order to maintain this catch when the stock is small, it is necessary to fish the available year classes so heavily that they are totally depleted before they reach spawning age. The positive minimum spawning stock when $h^* > 7$ consists of 7 year old fish which are not yet available for the fishery.

To sum up, it appears that the difference in the profitability of the two strategies considered here is not very great. Considering that the fluctuations in the abundance of the Arcto-Norwegian cod are probably greater than is the case for most other demersal stocks, this is likely to be all the more true for other demersal stocks. Similar simulations of the Icelandic cod stock, which fluctuates much less, confirm this; here the difference between the two strategies was even less than for the Arcto-Norwegian stock.

As a final remark, it must be emphasized that both of the strategies discussed here are somewhat *ad hoc* and neither is likely to be the optimal one. In Hannesson (1975) it is shown that periodic fishing in which a stock is fished intensively for one or a few years at a time and then left to recover yields a greater profit than sustained fishing with a constant fishing mortality. The practical constraints facing each particular fishery are likely not to allow such extensive fluctuations in catches and effort; the mobility of national fleets between stocks belonging to different sovereign states is limited, to say the least. Nevertheless,

Table 4.3 Zero cost per unit of fishing mortality. Kinked revenue curve ($R = Y$ for $Y < Q$, $R = Q$ for $Y > Q$).

h^*	4	5	6	7	8	9
Constant fishing mortality						
F	0.176	0.200	0.228	0.262	0.323	0.380
π	973.9	1047.0	1118.1	1179.9	1221.9	1227.5
Y_{av}	1005.9	1095.9	1184.3	1261.9	1328.5	1360.9
Y_{max}	1123.8	1272.4	1432.3	1606.9	1797.8	1931.3
Y_{min}	887.9	919.4	936.3	917.0	859.2	790.7
X_{min}	2983	2987	3076	3273	3416	3878
Constant annual catch						
Q	1018.0	1118.0	1224.0	1325.0	1422.0	1464.0
π	1018.0	1118.0	1224.0	1325.0	1422.0	1464.0
F_{av}	0.209	0.255	0.351	3.973	4.530	4.259
F_{max}	0.240	0.315	0.487	22.855	26.728	26.772
F_{min}	0.181	0.206	0.252	0.422	0.408	0.391
X_{min}	2227	1936	1433	0	1433	2953

h^* = age at first capture (knife-edge selectivity); F = fishing mortality; π = profit per year; Y_{av} = average catch per year; Y_{max} = maximum catch per year; Y_{min} = minimum catch per year; X_{min} = minimum size of the spawning stock; Q = annual constant catch; F_{av} = average fishing mortality per year; F_{max} = maximum fishing mortality per year; F_{min} = minimum fishing mortality per year.

we may conclude that some variation over time in catches and effort (fishing mortality) is likely to be better than keeping one of these constant, but by how much they ought to be varied depends on the specific circumstances in each particular fishery.

4.3 Optimum capacity with random catch quotas

Many fisheries are based on one or two year classes of fish. This is particularly true of species with a short lifespan (1 or 2 years), but also of species where the fishery is based on mature year classes returning to their spawning grounds (salmon, capelin). Given that the size of the year classes fluctuates for environmental reasons, a fishery based on a single year class can be stabilized only at a level sustained by the smallest year class. This level could be so low (possibly zero) as to make the stable catch option an unattractive one. Letting the catch fluctuate with the stock raises the question of how large a catching and processing capacity it is worth while to invest in, to make the best possible use of the catch potential. This is the question to be investigated in this section. Two cases will be examined, purely random total allowable catches, and cases where the future allowable catch depends on how much is being taken in the current season or year. The first of these cases was considered briefly in Chapter 2. What we offer here is a more formal analysis, considering explicitly the cost of fishing capacity and the probability distribution of the allowable catch.

4.3.1 *The case of a purely random total allowable catch*

The relationship between the size of the spawning stock and the size of the returning year class is usually not known with any great certainty. The management of fisheries based on spawning migration often follows the rule that a minimum spawning stock should be spared, in order to ensure future recruitment. This may in fact be the most sensible strategy, given the limited knowledge about the stock – recruitment relationship. If the harvestable biomass varies randomly because of environmental factors, the total allowable catch will also vary randomly under this management strategy. This approach is even more relevant when spawning takes place prior to fishing, as the future recruitment then is independent of how much is being fished.

As mentioned above (Section 4.2.3), environmental conditions sometimes appear to follow cyclical fluctuations of some regularity. This implies that the amount of fish that can be taken each year will fluctuate cyclically, provided that it is not possible or desirable to stabilize the catch. The probability distribution of the total allowable catch (Q) will then change over time. We shall, however, ignore this time-dependence, with the justification that such cycles are likely to be considerably shorter than the lifespan of the durable equipment that is needed in the fishery. The probability distribution of Q will therefore be regarded as time-invariant. We let the probability density of Q be given by the function $f(Q)$, so that the cumulative distribution function of Q becomes

$$F(\bar{Q}) = \int_0^{\bar{Q}} f(Q)dQ \tag{4.10}$$

The probability density function and the distribution function are illustrated in Fig. 4.5. The probability density function can be thought of as a frequency distribution of observed total allowable catches over a long time. The cumulative distribution function shows the probability that the total allowable catch will be less than or equal to the value $\bar{Q}$. Hence $F(Q)$ can only attain values between 0 and 1, including both of these. Denoting the upper and lower limit to the total allowable catch by Q_{max} and Q_{min}, we have

$$F(Q_{min}) = 0, \quad 0 < F(Q_{min} < Q < Q_{max}) < 1, \quad F(Q_{max}) = 1$$

It may be convenient for computational purposes to set $Q_{min} = 0$ and $Q_{max} = \infty$.

As discussed in Chapter 2, fishing effort is composed of various components which to some extent can replace one another. A fishing boat may be operated for

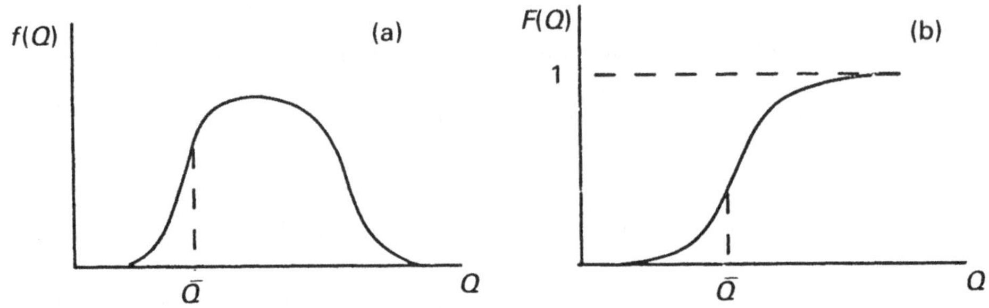

Fig. 4.5 A random total allowable catch: (a) probability density; (b) cumulative probability distribution.

a shorter period of time but maintain its catch rate by using a larger crew and more fishing gear. Here we shall, for simplicity, ignore this possibility and assume that fishing effort is proportional to the number of boats being used. We shall, furthermore, assume that the fleet consists of identical vessels (or can be meaningfully represented by a number of standardized vessels). Hence the actual effort will be proportional to the number of boats used (per unit of time), while the effort capacity will be proportional to the number of boats available. Since optimal investment is the issue we are concerned with here, we let the fleet size be represented by the amount of money invested in the fleet, denoted by K. We will not consider explicitly investment in processing capacity, but the analysis can easily be extended to cover this case.

Determination of optimum fleet capacity

Since the probability distribution of the total allowable catch is independent of time, it obviously makes sense to attain the optimum fleet capacity as quickly as is economically prudent and to maintain it ever after, unless some economic or technological parameters change. If the fleet is too large to begin with, the boats should not be replaced until the long-term optimum capacity has been attained, while the fleet should be enlarged as quickly as possible if it is too small initially, unless too rapid investment entails excessive costs. Formally it is easiest to deal with the problem where the fleet has to be built from scratch; the long term solution will in any case be the same regardless of the starting point. We assume that the cost of a boat is constant and independent of how many boats are built. The fleet should therefore be built as rapidly as possible, and we assume that this can be done in one time period (year). Once the fleet is in place, the profit per year will be $R - \delta K$, where R is the revenue net of operating costs and δ is the depreciation and maintenance cost per unit of capital invested in the fleet. Maximization of the present value of profits (rents, if the price is constant) can then be formalized as

$$\max \sum_{0}^{\infty} [ER(K) - \delta K]/(1 + r)^{t} - K \tag{4.11}$$

where ER is the expected revenue net of operating costs and r is the discount rate.

Since the probability distribution of the total allowable catch is time-invariant, the expected revenue will be the same every year. If, on the other hand, it were possible to predict the cyclical fluctuations in the total allowable catch, the expected revenue would change over time. The optimal size of the fleet would fluctuate cyclically, but less so than the expected revenue because of the durability of fishing boats. The renewal of the fleet would be accelerated when good years are predicted and delayed in the opposite case. We will not attempt a formal analysis of this here.

Solving the infinite series in (4.11) and multiplying by r provides an expression equivalent to (4.11) and a more useful one for our purposes:

$$\max ER(K) + (r + \delta)K \tag{4.12}$$

This expression has a straightforward common-sense interpretation. What we are maximizing is the annual expected revenue of the fishery, net of all costs. $R(K)$ is defined as the revenue net of operating costs, while $(r + \delta)K$ is the annualized capital cost, which consists of two components, the alternative cost of capital (rK), and the real depreciation and maintenance (δK).

The optimum solution to this problem is obtained by differentiating (4.12) with respect to K and setting the result equal to zero. This implies

$$ER'(K) = r + \delta \tag{4.13}$$

with a prime (′) denoting the first derivative. This has a straightforward interpretation. On the left-hand side we have the expected marginal revenue of capital per year or, in other words, the expected revenue of investing in an additional unit of production equipment (boats). On the right-hand side we have the annualized cost of a unit of capital, which amounts to the rate of depreciation (including maintenance) plus the opportunity cost of capital. Hence we should invest in boats up to the point where the expected annual revenue of one additional boat is equal to the annualized cost of one additional boat.

Catch capacity and the total allowable catch

To see more clearly what this amounts to, we need to develop an expression for $R(K)$. To do so we must consider how the annual catch is related to the total allowable catch and the capacity of the fleet. We shall continue to make use of the two special cases of the production function in the fishery (Equations (4.1a,b)). What we need to take into account now is that it will not always be necessary to use all of the available fleet capacity. How much of the fleet capacity will be needed depends on the size of the fishable stock and the size of the spawning

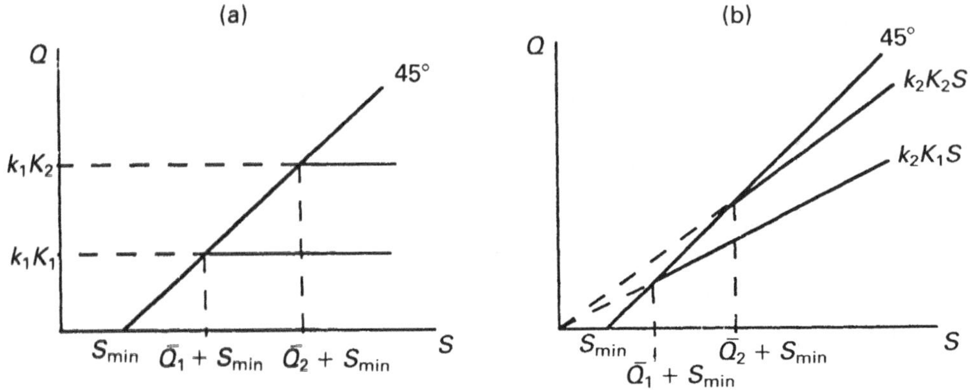

Fig. 4.6 Allowable catch ($S - S_{min}$) and fishing capacity: (a) constant catch per boat; (b) catch per boat proportional to stock (S).

stock to be left behind. Let the latter be denoted by S_{min}, so that the total allowable catch will be $Q = S - S_{min}$. The capacity limit (Q^*) of the fleet, in the two cases considered, is

$$Q^* = k_1K \tag{4.14a}$$

$$Q^* = k_2KS \tag{4.14b}$$

where K replaces effort in our earlier notation, for reasons already given.

Figure 4.6 shows how the capacity limit and the total allowable catch vary with the size of the stock. No catch will be permitted unless the stock exceeds the minimum escapement limit, but everything in excess of this can be taken, so the relationship between the stock size and the total allowable catch is a straight line with a slope of 1. If Equation (4.14a) holds, the capacity limit will be a horizontal line the location of which depends on K. Two such lines are shown in part (a) of Fig. 4.6. If Equation (4.14b) is valid, the capacity limit rises with the stock, as shown by the lines k_2K_1S and k_2K_2S in part (b) of Fig. 4.6, the slope of the line depending on the total capacity (K) of the fleet.

From Fig. 4.6 we see that the available fleet capacity (K) sets a limit to how much of the total allowable catch can be taken. In case (b) this limit is elastic, however. Even if all the total allowable catch cannot be taken once the total allowable catch exceeds the critical level $\bar{Q}$(determined by the intersection of the line representing Equation (4.14b) with the line showing the total allowable catch), the catch will increase with the stock as the latter increases beyond $\bar{Q} + S_{min}$. For the assumed management policy $Q = S - S_{min}$, we can write (4.14b) as

$$Q^* = k_2K(Q + S_{min}) \tag{4.14b'}$$

The actual catch in any given period will be the minimum of what is being allowed and what a fully utilized fleet can possibly take. In symbols:

$$Y = \min(Q, Q^*) \tag{4.15}$$

This is illustrated by the solid lines in Fig. 4.6.

It may be noted that these results are also valid for different policies for setting the total allowable catch, such as for example the policy where the total allowable catch is some fraction k_3 of the stock ($Q = k_3S$), in which case the fishing mortality would be held constant. The total allowable catch would then be given by a straight line with a slope k_3, starting at the origin. There is one exception, however, namely when the catch per unit of effort is proportional to the stock, as in part (b) of Fig. 4.6. In this case there is one level of capacity that will always be fully utilized; that is, the one which produces a line coinciding with the straight line with a slope k_3 showing how the total allowable catch varies with the stock. All other capacity levels will be either too high or too low. In this case, which involves a special rule for setting the total allowable catch and a special case for the production function, there is no capacity choice to be made independently of how the total allowable catch is decided; once it has been decided what fraction of the stock to harvest, the optimal catch capacity has been decided as well. This capacity will always be fully utilized. Note, however, that operating costs that are related to capacity actually used implies that operating cost per unit of fish caught is inversely related to the stock level (see Equation (4.16′b) below). There is, therefore, a critical stock level at which the operating cost just gets covered and below which it is unprofitable to deplete the stock. If this critical level is greater than the smallest stock ever left behind according to the harvest rule, we have in fact a case similar to the rule $S - S_{\text{min}}$

Revenue, price and cost

The revenue net of operating costs (R) of any given catch is equal to the catch volume (Y), multiplied by its price (p) less operating costs per unit caught (c_y). The price may depend on the catch volume, while the operating cost per unit of fish caught will depend on the size of the fish stock, unless the catch per unit of capital (K) invested in the fleet is independent of the size of the stock. This may be seen as follows. Define K_a as the necessary effort when the total allowable catch is less than what the fleet can take with full capacity utilization. Let operating costs (c) be proportional to the amount of effort actually used. The operating cost per unit of fish caught (c_y) will then be

$$c_y = cK_a/Y \tag{4.16}$$

From Equation (4.1) we can find the amount of effort actually used:

$$K_a = Y/k_1 \tag{4.17a}$$

$$K_a = Y/k_2S \tag{4.17b}$$

Substituting this in (4.16) we get

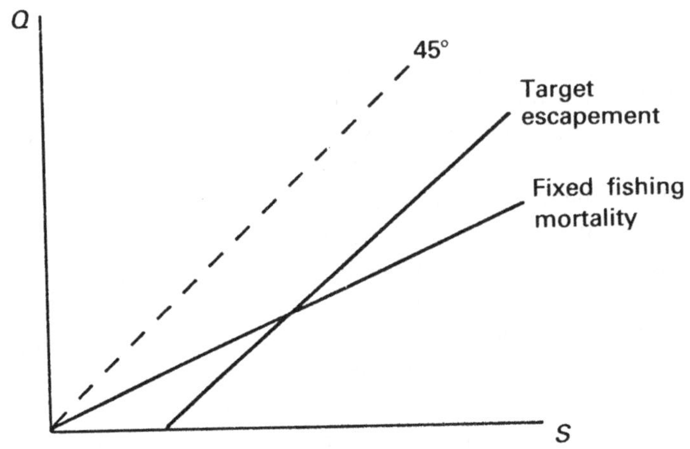

Fig. 4.7 Relationship between total allowable catch (Q) and stock size (S) for two common policy rules.

$$c_y = c/k_1 \qquad (4.16'a)$$

$$c_y = c/k_2S \qquad (4.16'b)$$

Suppose that the total allowable catch (Q) is related to the size of the stock (S) by a rule saying that Q increases monotonically as S increases beyond a certain level S_{min}:

$$Q = f(S), \quad S > S_{min}$$

The monotonicity of the function $f(.)$ implies that S can be written as a function of Q, for $S > S_{min}$

$$S = f^{-1}(Q), \quad S > S_{min}$$

where $f^{-1}(.)$ is called the inverse function of $f(.)$

Figure 4.7 shows two common policy rules, $Q = S - S_{min}$ (the target escapement rule) and $Q = k_3S$ (the fixed fishing mortality rule), both of which we have already discussed. For these two rules we get

$f(S) = S - S_{min}$ and $f^{-1}(Q) = Q + S_{min}$ (target escapement)

$f(S) = k_3S$ and $f^{-1}(Q) = Q/k_3$, $S_{min} = 0$ (fixed fishing mortality)

For these rules, Equation (4.16′b) becomes:

$$c_y = c/k_2(Q + S_{min}) \qquad (4.16'b1)$$

$$c_y = ck_3/k_2Q \qquad (4.16'b2)$$

From this it is clear that c_y will depend on Q and decrease as Q increases, except in the special case of Equation (4.16′a), so we write $c_y(Q)$. The value of a unit of fish caught (before subtracting capital costs) will be

$$p(Q) - c_y(Q) \quad \text{if} \quad Q < Q^*$$

$$p(Q^*) - c_y(Q) \quad \text{if} \quad Q > Q^*$$

Here it may be noted that the price depends on the catch actually taken (Q^*), while the unit cost depends on the total allowable catch (Q) through the latter's dependence on the stock.

Optimum capacity again

We are now equipped to find the expected revenue per year. This involves multiplying all possible catch values with the probability that they will occur and sum over all catch values. Since we are treating the catch as a continuous variable we must do the summation in infinitely small steps as it were, which amounts to calculating an integral instead of a sum:

$$ER(K) = \int_0^{\bar{Q}(K)} [p(Q) - c_y(Q)]Qf(Q)dQ + \int_{\bar{Q}(K)}^{\infty} [p(Q^*(K,Q)) - c_y(Q)]Q^*(K,Q)f(Q)dQ \tag{4.18}$$

The 'sum' expressing ER, the expected revenue, consists of two parts. If the total allowable catch (Q) is less than the minimum catch taken when all the available fleet capacity is used ($\bar{Q}$), the actual catch will be equal to the total allowable catch. This is the first part of the 'sum' on the right hand side of Equation (4.18). The value of the catch is obtained by multiplying by the price less operating cost per unit caught. The expected value of the catches within this range (0, $\bar{Q}$) is given by multiplying by the probability density function and summing (integrating) over the relevant interval of catches.

The latter term takes care of the catches obtained when the total allowable catch is greater than the fleet capacity (Q^*). The actual catch will then be equal to the fleet capacity, and again we obtain the value of the catch by multiplying by the price net of operating cost per unit. The expected value is obtained by multiplying the value of the catch by the probability density of the total allowable catch (Q) corresponding to the actual catch being taken and summing (integrating) over the relevant interval of the total allowable catch ($\bar{Q}$, ∞). Note that the catch taken by a fully utilized fleet depends on the capacity of the fleet (K) and possibly also on the total allowable catch (see 4.14b′).

To apply Equation (4.13) we must calculate ER', the first derivative of ER, or, in non-mathematical language, the increase in the catch value on the average (expected catch) when the fishing capacity is increased by a small amount. The term ER' can be calculated by the rule of differentiation of integrals:

$$ER'(K) = \int_{\bar{Q}(K)}^{\infty} \{p'(Q^*)Q^*(K,Q) + [p(Q^*) - c_y(Q)]\}Q^*_k(K,Q)f(Q)dQ \quad (4.19)$$

where subscript k denotes partial derivative with respect to K.

A noteworthy feature of this equation is that the expected incremental revenue resulting from a marginal increase in fleet capacity depends on what happens to the value of the catch when the total allowable catch is greater than the capacity of the fleet ($Q > Q^*$). When the catch is less than the capacity of the fleet, the catch value will not be affected at all by an increase in fishing capacity. This is why the 'sum' on the right-hand side of (4.19) begins at $\bar{Q}$, the catch level at which all the fleet capacity is needed.

Note that the first term in brackets $[p'(Q^*)Q^*(K,Q)]$ is not relevant from the social point of view, unless the fish is exported, in which case consumer surplus does not have to be taken into account. Substituting consumer surplus for revenue in Equation (4.18), we would get, instead of the terms $p(Q)$ and $p(Q^*)$:

$$\int_0^Q p(x)dx \quad \text{and} \quad \int_0^{Q^*} p(x)dx$$

When taking derivatives with respect to K, we would get $[p(Q^*) - c_y]Q^*_k$ instead of $[\cdot]Q^*_k$ in Equation (4.19). As explained in Chapter 2, maximizing revenue at a constant price equal to the price at the optimum gives the same answer as maximizing consumer surplus.

Let us look at a few special cases of (4.19) of increasing complexity. Suppose, first, that the price of fish and the catch per unit of effort are both constant. By Equation (4.14a), $Q^*_k(K,Q) = k_1$, while $c_y = c/k_1$ by Equation (4.16a), so we have

$$ER'(K) = \int_{\bar{Q}(K)}^{\infty} [pk_1 - c]f(Q)dQ$$

which, by Equation (4.10) and the fact that $F(0) = 0$ and $F(\infty) = 1$ can be written as

$$ER'(K) = [1 - F(k_1K)](pk_1 - c) \quad (4.19')$$

so Equation (4.13) can be written as

$$[1 - F(k_1K^o)] = (\delta + r)/(pk_1 - c) \quad (4.13')$$

This equation implicitly determines the optimal amount of capital (K^o) to be invested. The solution is illustrated in Fig. 4.8. We see that it is never optimal to invest in a fleet capacity that can take the maximum allowable catch, unless the cost of capital ($\delta + r$) is zero. A higher cost or a lower productivity of boats would lower k_1, the amount of fish caught by one unit of money invested in boats. This would lower the optimum capacity and the capacity limit ($\bar{Q}$, which here is equal to the constant Q^*). The same occurs if the price of fish (p) falls, or the

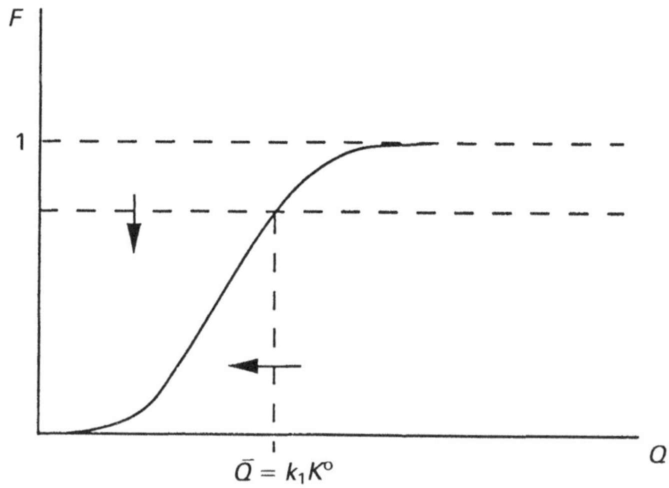

Fig. 4.8 Cumulative probability distribution for the total allowable catch (Q), and the optimum capacity (K^o). Arrows indicate the effect of a higher cost or a lower price or productivity.

operating cost of boats (c) or the cost of capital ($\delta + r$) increases. This is indicated by the arrows in the figure.

Then let us look at the case where the catch per boat is proportional to the size of the fish stock. If the fixed fishing mortality rule is being followed, the optimum fishing capacity will be determined once the optimum fishing mortality is determined, as explained above. Let us, therefore, look at the other rule where a target escapement policy is being followed. By Equation (4.14b′), $Q_k^*(K,Q) = k_2(Q + S_{min})$, and by Equation (4.16′b1), $c_y = c/k_2(Q + S_{min})$. Equation (4.19) becomes

$$ER'(K) = \int_{\bar{Q}(K)}^{\infty} [pk_2(Q + S_{min}) - c]f(Q)dQ \qquad (4.19'')$$

The difference between this and (4.19′) is that the catch of an additional unit of capital is not constant but varies with the size of the stock ($Q + S_{min}$). We must, therefore, find the expected catch of an additional boat by multiplying the catch per boat for each given stock level with the probability that this particular stock level will occur and sum over all stock levels above $\bar{Q} + S_{min}$.

If the price of fish depends on the quantity caught, we get, for the two cases of the catch per unit of capital (effort):

$$ER'(K) = [1 - F(k_1K)]\{[p'(k_1K)k_1K + p(k_1K)]k_1 - c\} \qquad (4.19''')$$

$$ER'(K) = \int_{\bar{Q}}^{\infty} \{[p'(Q^*)k_2K(Q + S_{min}) + p(Q^*)]k_2(Q + S_{min}) - c\}f(Q)dQ \qquad (4.19'''')$$

Table 4.4 Norwegian catches of Barents Sea capelin 1970–1990.

Year	Catch (Q) '000 t	ln Q
1970	1131	7.1808
1971	1393	7.2392
1972	1592	7.3728
1973	1336	7.1974
1974	1149	7.0466
1975	1417	7.2563
1976	2545	7.8419
1977	2940	7.9862
1978	1894	7.5464
1979	1783	7.4861
1980	1648	7.4073
1981	2006	7.6039
1982	1760	7.4731
1983	2304	7.7424
1984	1461	7.2869
1985	851	6.7464
1986	123	4.8122
1987	0	$-\infty$
1988	0	$-\infty$
1989	0	$-\infty$
1990	0	$-\infty$

As already stated, the term $p'(.)$ is not relevant from a social point of view unless the fish is being exported.

An example of the Barents Sea Capelin

Like many other salmonid fishes, most capelin die after spawning. The fishery is based on two mature or nearly mature year classes. The variations in year class strength, which are substantial, are therefore reflected in fluctuations of the fishable stock.

For the last 15–20 years the Barents Sea capelin has been managed on the principle that a sufficiently large spawning stock should be left behind to secure the recruitment to the stock. This implies that the catches fluctuate even more than the fishable stock. Table 4.4 shows the total Norwegian catches of capelin since 1970 (from 1977 Norway and the Soviet Union managed this stock jointly and shared the total allowable catch, with 60 per cent being allocated to Norway). According to the table, the catches have varied between zero and nearly 3 million t over the said period. Since this stock has been the most important resource for the highly specialized fleet of purse-seiners in most years since about 1970, the capacity utilization of this fleet has been very uneven. As the scope of evening out such fluctuations is limited in a fishery that is based on so few year classes, the

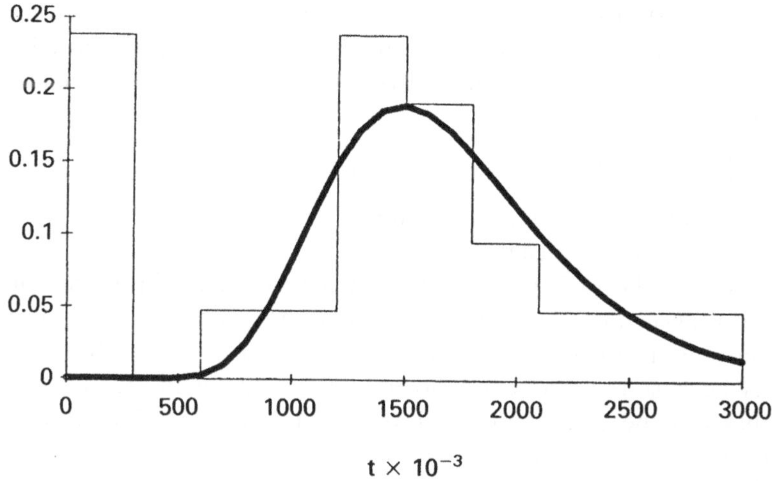

Fig. 4.9 Norwegian catches of Barents Sea capelin. Frequency and probability density.

investment in fleet capacity should be determined on the basis of expected variations in the catch. Neglecting to do so may cause over-investment in boats that will seldom be needed.

The figures in Table 4.4 indicate that the probability distribution of the total allowable catch is not the same every year. Over the last 5 years the catch has been either zero or insignificant, and in previous years the catch profile looks more like waves than random variations from year to year. It therefore appears that the probability of a large total allowable catch in a particular year will be greater if the total allowable catch was high the previous year, and vice versa. The autonomous probability distribution that we assumed above therefore does not quite fit the facts of the Barents Sea capelin fishery. We shall nevertheless use an autonomous probability distribution function as an illustration, with the justification that a 'boom' or 'dearth' of 5 years or so is still considerably shorter than the lifetime of a fishing vessel (the capelin fishery in the Barents Sea was resumed in 1991). Hence there is limited scope to adapt the capacity of the fleet to the time pattern in the fluctuations of the total allowable catch.

Figure 4.9 shows the frequency distribution of the catches in the period 1970–1990 as columns. Each column covers an interval of 300 000 t and shows the number of years with a catch falling within each interval, divided by the total number of years. This distribution is 'bimodal'; catches in the interval 0–300 000 t are as common as catches in the interval 1 200 000–1 500 000 t. The catches in the first interval are zero with one exception, a catch of 123 000 t. The distribution of the remaining catches is skewed, with a long right-hand tail.

Figure 4.9 suggests that we distinguish between very small catches (catches in the interval 0–300 000 t) and larger catches. Provided that the total allowable catch exceeds 300 000 t, the probability distribution seems to fit the lognormal distribution tolerably well. The curve in Fig. 4.9 shows a lognormal distribution fitted to the catch data for the years 1970–1985. The lognormal probability density function is

$$f(Q) = [\sqrt{2\pi}\sigma Q]^{-1}\exp[-(\ln Q - \mu)^2/2\sigma^2] \tag{4.20}$$

and the estimated parameter values are $\mu_x = 7.4009$ and $\sigma_x = 0.3091$, $x = \ln Q$, Q being measured in thousands of tonnes.

In order to make (4.20) comparable to the frequency histogram of the actual catches, it is necessary to multiply each point on $f(Q)$ by the width of the columns in the histogram (300) and divide by $(1 - 5/21)$. The latter adjustment is necessary because the frequency shown by the columns in Fig. 4.9 was calculated on the basis of the catches in all 21 years, while the last 5 years were excluded when fitting the lognormal density curve.

We shall now illustrate, with a simple example based on some stylized facts about the Norwegian purse-seining fleet, how the model developed above can be used to calculate the optimum long-term fleet capacity. There is some evidence that the catch per unit of effort of the purse-seine fleet is nearly constant and independent of the size of the capelin stock. Hence we shall use Equation (4.13′), which we reproduce for later reference:

$$[1 - F(k_1K^o)] = (r + \delta)/(pk_1 - c) \tag{4.13′}$$

This equation determines the optimal value of the capital stock (K^o) implicitly. Before showing how, we need to determine the parameters k_1, r, δ and p. The discount rate, r, should reflect the alternative rate of return on investment, which we put at 5 per cent for illustrative purposes. The depreciation rate is the inverse of the lifetime of the equipment in question. The economic lifetime of equipment necessary for fishing purposes varies a great deal from one type of equipment to another. Fish finders and other electronic devices last for a few years while the vessel hull may last for decades if adequately maintained.

The assumed lifetime of capital equipment identified in the Norwegian investigations on the profitability of fishing vessels (vessels, engines, electronic equipment, hydraulic equipment, freezing equipment) ranges from 7 years for freezing equipment to 30 years for vessel hulls. As the vessel hull represents the largest part of the value of a fishing boat, including all durable equipment, it is not unreasonable to assume an average lifetime of capital of 20 years, which would give a depreciation rate of 5 per cent. To this must be added the cost of necessary maintenance of the equipment. As the optimum solution turns out to be quite sensitive to the capital cost, we shall make the cautious assumption that depreciation and maintenance amount to 5 per cent of the cost of a new vessel. This gives a capital cost $(r + \delta)$ of 0.1.

In recent years the price of capelin has been about 600 Norwegian kroner per tonne. The operating cost, including labour cost, probably is about one-half of this or more, so we set the net price ($p - c/k_1$) of capelin at 300 kroner per tonne. This leaves k_1 to be determined. This parameter shows the amount of capelin (in tonnes) caught by one unit of capital invested. A large vessel can catch roughly 25 000 t of capelin each year and costs about 50 million kroner, which gives $k_1 = 25/50\,000$, k_1 being measured as tonnes per krone.

We can now determine the right-hand side of Equation (4.13′):

$$(r + \delta)/(p - c/k_1)k_1 = 0.1/300(25/50\,000) = 0.6667$$

It remains to find the value of K which makes the left-hand side of (4.13′) equal to 0.6667. Now, $F(k_1K^o)$ is the probability that the total allowable catch will be less than the catch capacity of the fleet, k_1K^o. The fact that we have assumed that the total allowable catch is lognormally distributed, once the catch exceeds 300 000 t, makes our task relatively easy. Note that $k_1K^o = Q^*$ (see Equation (4.14a)), the catch at which all available capacity must be fully used. The probability that the catch will be less than this is equal to the probability that the catch will be below 300 000 t, which we have set at 5/21, plus the probability that the catch will be less than Q^* but greater than 300 000 t. The latter is equal to the probability of $Q < Q^*$, as determined by Equation (4.20), times the probability that the catch will exceed 300 000 t. This gives

$$F(Q^*) = 5/21 + F(z^*)(1 - 5/21)$$

where $F(z^*)$ is the probability that the normalized variable, $z = (\ln Q - \mu_x)/\sigma_x$, will be less than $z^* = (\ln Q^* - \mu_x)/\sigma_x$. This can also be written as

$$F(z^*) = [F(Q^*) - 5/21]/(1 - 5/21) \tag{4.21}$$

From (4.13′) we can calculate $F(Q^*) = 0.3333$, and from (4.21) we can find $F(z^*) = 0.1250$. Standard probability tables provide the values of z for various probabilities. From this we find $z^* = -1.15$. The corresponding value of Q^* we find by running the normalization procedure 'backwards', that is, $\ln Q^* = \mu_x + \sigma_x z^*$, which gives $Q^* = 1148$. The optimum capacity thus amounts to an annual catch of 1 148 000 t, which is substantially less than the average catch.

The optimum investment can be found by dividing by k_1, the number of tonnes produced by one krone invested. This gives $K^o = 1\,148\,000/(25/50\,000) = 2296$ million kroner.

The optimum solution is very sensitive to moderate changes in the economic parameters. Suppose, for example, that we increase the capital cost ($r + \delta$) to 0.15, which is by no means unrealistic. The right-hand side of (4.13′) would become $0.15/300(25/50\,000) = 1$. This implies $1 - F(Q^*) = 1$, or $F(Q^*) = 0$, which implies $\bar{Q} = Q_{\min} = 0$. In this case it would not be worth while to invest in any fishing capacity at all. The same occurs if, with $r + \delta = 0.1$, the net price is reduced to 200. The reason is the rather high volatility of the annual catch and the significant probability of no catches at all being allowed.

Note that these results depend on the implicit assumption that there is no use for the fishing vessels outside the fishery under consideration. If there are other fisheries in which the vessels can participate, these would have to be taken into consideration explicitly. The single stock analysis provided here may nevertheless be appropriate in a multifishery setting. For a fleet exploiting several stocks sequentially, it will often be true that one particular stock demands a greater catch capacity than the rest. Even if the profitability of one additional vessel depends on the probability that additional capacity will be needed in the fishery as a whole, the probability of needing it for the stock normally demanding the greatest capacity is likely to be decisive.

4.3.2 *A stochastic recruitment model*

In this section we shall consider a dynamic model in which the development of the stock depends on a stochastic growth function (a function with a random element). We will use the Ricker recruitment function for this purpose and apply it to the Barents Sea capelin. Since the capelin fishery is based on two mature, or nearly mature, year classes of fish, the relationship between the size of the spawning stock and the recruitment of a new year class to the fishery is particularly important for a dynamic model of the stock. If no significant relationship of this kind can be found, the available amount of fish must be regarded as either purely random, as in the previous section, or possibly time-dependent but stock-independent, owing to environmental 'waves' that last for several years.

The Ricker recruitment function (Equation (3.15)) is

$$N = X \exp[a(1 - X/S^*)] \tag{4.22}$$

where N is recruitment in numbers and X is the weight of the spawning stock.

For the capelin stock there are data available for the size of the spawning stock in the period 1973–87 and for the size of the resulting year class as 2 year old fish (see Table 4.5). Taking logarithms of Equation (4.22) and using subscripts to indicate the year to which the data apply, we get

$$\ln(N_{t+2}/X_t) = a - bX_t + u_t \tag{4.23}$$

where $b = a/S^*$, and u is a normally distributed error term with an expected value of zero and a constant variance. Note that in this stochastic specification it is no longer possible to interpret S^* as the equilibrium stock size, in the sense that the expected recruitment will be S^*, given that the spawning stock is S^*. The reason for this is that the error term is exponential (e^u), which implies that $EN(S^*) > S^*$. If we want to specify the recruitment function in such a way that $EN(S^*) = S^*$, or more generally, that the expected recruitment for any given size of the spawning stock will be equal to that given by a deterministic recruitment curve with the same parameters (that is, $u = 0$ always), we would have to subtract the term $\sigma^2/2$ in the exponential term (see, for example, Lindgren 1968).

Table 4.5 Spawning stock and recruitment of Barents Sea capelin 1973–1987.

Year (t)	Spawning stock ($\times 10^3$ t) year t	Recruits (10^9) year $t+2$	Assumed weight of recruits ($\times 10^3$ t)
1973	783	650	4501
1974	288	435	3150
1975	268	380	2756
1976	1328	708	2043
1977	1243	607	4045
1978	685	356	1691
1979	320	377	2010
1980	228	281	1182
1981	2023	390	1111
1982	720	341	310
1983	215	102	42
1984	302	62	1
1985	–	–	–
1986	36	29	<1
1987	16	18	30

Estimating the parameters a and b from the data provided in Table 4.5 gave the following results:

$$\ln(N/X) = \underset{(0.59)}{0.16} - \underset{(-4.94)}{0.00089X} \quad R^2 = 0.59 \quad F = 24.36$$

Numbers in parenthesis show t – values. The estimation method used corrects for autocorrelation.

The recruitment function, using the estimated parameter values, is shown in Fig. 4.10 together with the observed recruitment. Three versions of the recruitment function are shown in the figure. The one in the middle corresponds to $u = 0$, while the upper and lower ones correspond to $u = \pm 2\sigma$. According to statistical theory, the actual value of u should be within the interval of $\pm 2\sigma$ with 95 per cent probability. With one exception, the observed recruitment falls within the boundaries of the upper and lower recruitment curve. The figure gives the impression, however, that the estimated recruitment function may exaggerate the probability of exceptionally large recruitment emerging from a medium size spawning stock, while it may underestimate the probability of exceptionally small recruitment.

The recruitment curves in Fig. 4.10 show the relationship between the size of the spawning stock in tonnes and the number of recruits as two year old fish. What we need for our present purpose is a relationship between the size of the spawning stock in tonnes and the size of the fishable stock in tonnes. To get from recruitment in numbers to recruitment in terms of fishable biomass involves two

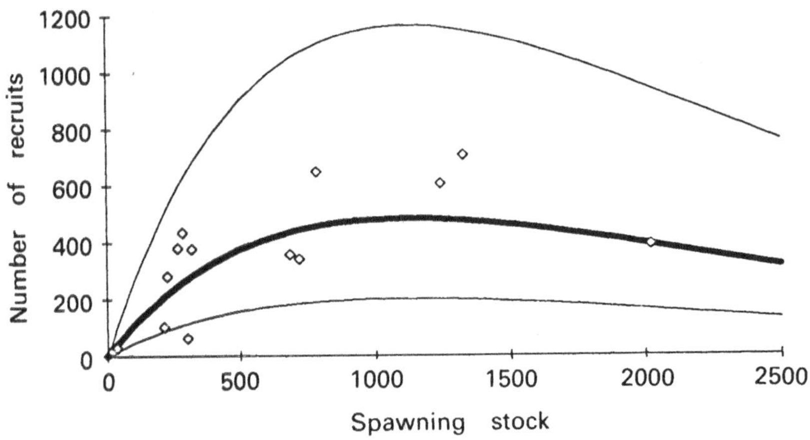

Fig. 4.10 Barents Sea capelin. Stochastic recruitment curves and observed recruitment. The curves are drawn for three alternative values of the stochastic variable u, $\pm2\sigma$ (upper and lower curve) and 0 (the 'thick' curve). Dots show actual recruitment (see Table 4.3).

steps. First, we must know the weight of each fish when it has reached the fishable age. Second, we must know how large a fraction of the 2 year old recruits survive to the fishable age.

The said relationships are complicated because at least two year classes are being fished each year, and the fishery takes place in two distinct seasons. During the summer season the fishery takes 2 and 3 year old capelin, while the winter season the following calendar year is based on the mature stock, consisting mainly of 4 year old fish. We shall, for simplicity, assume that all capelin are fished when they are four years old.

We then need to know the weight of 4 year old capelin and how large a fraction of a 2 year old year class survives to the age of 4. The weight at age varies a great deal and probably depends on the size of the year classes, besides environmental factors. Here we shall make the simplifying assumption that 4 year old capelin weigh 12 g. Similarly the mortality of capelin is likely to vary over time, but we shall assume that the instantaneous mortality rate is 0.4 on an annual basis. Using the above estimates of the parameters of the recruitment function, we arrive at the following relation between the spawning stock in year t and the fishable stock in year $t + 4$, measured in thousands of tonnes:

$$S_{t+4} = 12X_t \exp[0.16 - 0.00089X_t - 0.8 + u_t] \qquad (4.24)$$

Equation (4.24) is shown in Fig. 4.11. The three curves shown correspond to the three recruitment curves in Fig. 4.10. Also shown in Fig. 4.11 are assumed sizes of the recruited year classes as biomass of exploitable fish. These figures are shown in the last column of Table 4.5 and were calculated as the weight of 3 and 4 year old fish, as estimated by acoustic surveys taking place after the summer fishing season is over, plus the amount caught by the Norwegian fleet during the

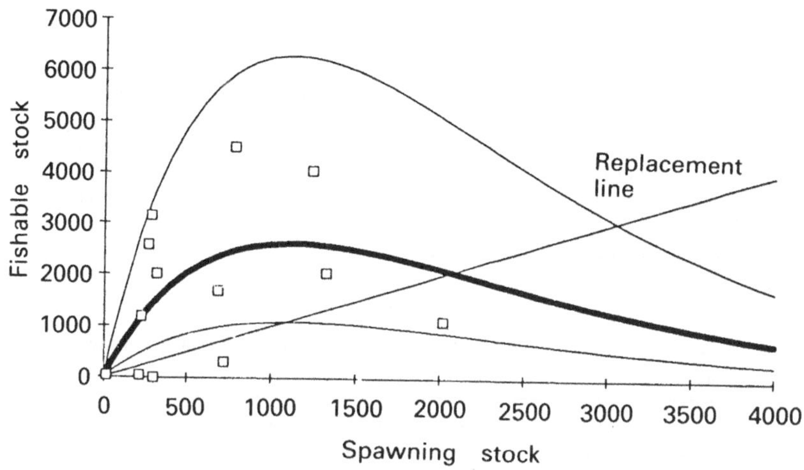

Fig. 4.11 Barents Sea capelin. Recruitment curve and estimated fishable stock. The three curves correspond to the curves in Fig. 4.10. the dots show assumed weight of the recruited year classes (see Table 4.5).

summer season. This is clearly an imprecise estimate, not least because we are counting two year classes instead of one, but is meant to provide an indication as to whether the assumed recruitment in terms of fishable biomass is wide of the mark.

Comparing the curves with the estimated size of the biomass of the recruited year classes, the recruitment function seems to overestimate the probability of exceptionally large recruitment and to underestimate the probability of exceptionally small recruitment. Despite these shortcomings, the estimated recruitment curve can still be used to illustrate how a dynamic model with a stochastic component works.

A discrete model

In order to perform computations with this model it needs to be made discrete. There are two steps to this process. First, we let all stock and recruitment values in the interval S_i, $S_i + \Delta S$ be represented by the mid-point of the interval. In the examples to follow, we set the beginning of the first interval at zero while ΔS is set equal to 200 (thousand tonnes), which implies that the lowest possible stock and recruitment level is 100 (thousand tonnes). The last interval has no upper limit, so as to include all possibilities, while being represented by the point $\Delta S/2$ above its lower limit.

Second, we let the values of the standard, normal variable $z = u/\sigma$ in the interval z_i, $z_i + \Delta z$ be represented by the mid point in the interval, except for the intervals at either end which extend from ± 2.9 to $\pm\infty$ but are represented by ± 3.0. The probability that the value of z will occur in the interval z_i, $z_i + \Delta z$ is easily calculated from probability tables. We set $\Delta z = 0.2$, which implies that z

Table 4.6 Probabilities of the standard normal variable z, represented by the mid-point in each interval.

z_i	Pr(z)	z_i	Pr(z)	z_i	Pr(z)	z_i	Pr(z)
3	0.0019	2.2	0.0072	1.4	0.0300	0.6	0.0665
2.8	0.0016	2	0.0108	1.2	0.0389	0.4	0.0736
2.6	0.0027	1.8	0.0159	1	0.0484	0.2	0.0781
2.4	0.0045	1.6	0.0222	0.8	0.0579	0	0.0796

Table 4.7 Probabilities of recruitments emerging from a spawning stock of 1100.

S_i	Pr(S)	S_i	Pr(S)	S_i	Pr(S)	S_i	Pr(S)
100	0.0000	1100	0.0267	2100	0.0736	3100	0.0736
300	0.0000	1300	0.0522	2300	0.0781	3300	0.0665
500	0.0000	1500	0.0389	2500	0.0796	3500	0.0000
700	0.0035	1700	0.0484	2700	0.0000	3700	0.0579
900	0.0144	1900	0.1244	2900	0.0781	3900	0.1841

can assume 21 discrete values ($-3, -2.8, \ldots, 2.8, 3$). For each particular value of z, the implied value of the stochastic variable u can be calculated from

$$u_i = \sigma z_i, \quad i = 1, \ldots, 31 \tag{4.25}$$

For any given spawning stock, X_j, all possible (31) values of u_i are substituted into the recruitment function and the corresponding recruitment calculated. The recruitment will occur in one of the intervals S_i, $S_i + \Delta S$, or in the uppermost interval S_1, ∞. The probability that the recruitment will occur in the interval S_i, $S_i + \Delta S$, given the spawning stock X_j, is given by the probability of u causing the recruitment to occur in this particular interval. Note that more than one value of u may cause the recruitment to occur in the same interval for a given size (X_j) of the spawning stock, in which case we must add the probabilities of all the values of u that cause the recruitment to occur in the same interval.

As an example, consider the probability matrix of z shown in Table 4.6. The probabilities for the negative values of z are the same, because the probability density of the standard, normal variable is symmetric.

Now look (Table 4.7) at the probability matrix for recruitments resulting from a spawning stock of 1100. Using Equations (4.24) and (4.25), it may be calculated that the values -3 and -2.8 of the standard, normal variable produce recruitments in the interval 600–800. The sum of probabilities of these two values of z are 0.0035, the probability of a recruitment of 700 shown in Table 4.7. Likewise, it can be verified that all values of $z \geqslant 1$ produce recruitments above 3800. The sum of probabilities of $z \geqslant 1$ is 0.1841, which is the probability of a recruitment of 3900 in Table 4.7.

Optimal capacity utilization

It now remains to maximize the expected present value of the fishery over a given time horizon. We shall assume an infinite time horizon, in order to avoid having to specify terminal conditions for the stock and to be able to focus on solution patterns that remain the same for a long period of time, given that no economic or technological parameters change. As in the purely random case above, there are two kinds of decisions to be taken; how much of a given fishing capacity to utilize in any given period, and how large a fishing capacity to build and to maintain. As in the purely random case, we shall assume that the fishing capacity will be maintained for ever once it has been acquired, but it is appropriate to note that this is a less realistic assumption here, since the probability distribution of the stock a year from now in this case depends on how large a spawning stock was left behind 3 years ago. Hence the fleet capacity can to some extent be adapted to the expected stock by not replacing worn out boats in years when the probability of a large stock emerging is small. The optimum capacity to be derived below is thus to be understood as an average around which the actual optimum capacity will vary, according to the status of the stock at each particular time.

We look first at the optimum utilization of a given fleet capacity. We shall consider only the case in which the catch per boat is constant and independent of the stock level and the price net of operating cost is constant. Optimum capacity utilization implies maximizing the expected revenue at a given price net of operating cost. This may be formalized as follows:

$$\max \sum_{1}^{\infty} (1 + r)^{-t}[ER(Y_t,\bar{K}_t)] \qquad (4.26)$$

where

$$R(Y_t,\bar{K}_t) = \min[(p - c_y)Q_t, (p - c_y)k_1\bar{K}] \qquad (4.27)$$

Q_t being the total allowable catch in year t.

The optimum capacity utilization may be found by applying Bellman's principle of dynamic programming. Define the optimal value function, V:

$$V_t(\bar{K},S_t) = \max_{Y_t < S_t}[R(Y_t,\bar{K}) + (1 + r)^{-4}EV_{t+4}(\bar{K},S_{t+4})] \qquad (4.28)$$

The discount factor is raised to the power of 4, because of the 4 year time lag from spawning until the new year class has reached fishable age.

Equation (4.28) defines the optimum capacity utilization, or catch (Y), as a function of the stock size for a given fleet, $Y^o(S;\bar{K})$. Because of the infinite time horizon, the optimum value function will be the same in period $t + 4$ as in period t for identical values of S. The optimum capacity utilization will therefore be the

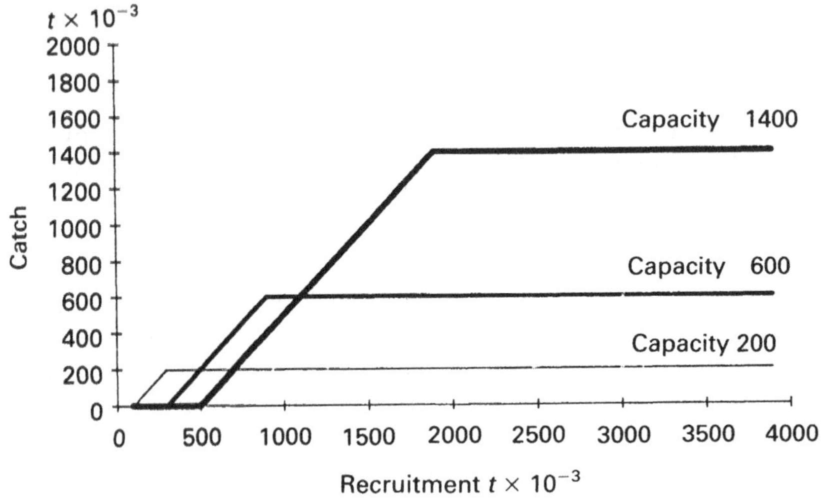

Fig. 4.12 Optimum capacity utilization.

same in all periods for any given value of S and may be found by the method of successive approximations (see, e.g., Dreyfus & Law 1977). This method involves setting $V_{t+4} = 0$ initially, maximizing V_t for all possible values of S, a substitution of the values $V_t(S_i)$ so found into $V_{t+4}(S_i)$ for all S_i, a maximization of V_t again, etc. The expected value of V_{t+4} for any given S_t is found by multiplying each value of $V(S_i)$ by its probability and summing over all possible values of $V(.)$. These probabilities can be found as follows. Subtract the optimal catch, Y^o, from the stock at the beginning of the harvesting period, S_t, to find the spawning stock $X_t = S_t - Y^o(S_t,\bar{K})$. For each given spawning stock, it is possible to calculate the probability that a stock of size S_i will emerge in the next period. One illustration was provided in Table 4.7 above, where the spawning stock was 1100. The table shows the probability of a stock emerging in the interval 0–200, 200–400, . . . , 3800+.

In Fig. 4.12 we show how the optimum catch varies with the size of the stock, for three different levels of fleet size (fishing capacity). The smallest possible fleet (200) will always be fully used, except when the stock is only 100 and would be wiped out if fished. At higher stock levels everything that it is possible to take will be taken. Since the capacity is small, there is no point in leaving a large spawning stock behind in order to increase the probability of a large recruitment, as only 200 (thousand tonnes) can be fished anyway. So, even if the stock is as small as 300, the capacity is fully used and a spawning stock of only 100 is left behind.

When the capacity is increased to 600, it begins to make sense not to use all of it when the stock is relatively small, but instead leave a large spawning stock behind in order to increase the probability of a large recruitment. We see that the capacity is not used at all unless the stock exceeds 300 (is 500 or more). This policy of not fishing the stock until it has reached a certain level, in order to improve the prospects for a large recruitment, is still more pronounced if the capacity is as large as 1400. Here the stock is left unexploited unless it exceeds 500 (is 700 or more). If the stock is in the interval 700–1900, a constant stock of 500 is left behind to breed, and the full capacity of the fleet is used only if the stock is 1900 or greater.

Optimum fleet capacity

It remains to consider the optimum size of the fleet. Consider a change ΔQ in the capacity of the fleet. Since the fleet is supposed to be maintained unchanged for ever once the optimal size has been attained, the cost of the said change will be

$$c_k \Delta Q + \delta c_k \Delta Q / r \quad (4.29)$$

where c_k is the cost of acquiring one unit of fishing capacity (the cost of building a boat, for example). The second term in this expression is the present value of the annual expenses necessary to cover depreciation and maintenance, in order to keep the fleet permanently at the new and higher level.

The benefit of a marginal increase in the fleet size is the resulting increase in the present value of revenue less operating cost, that is

$$(1 + r)^{-1}[EV(\bar{Q} + \Delta Q) - EV(\bar{Q})] \quad (4.30)$$

where the additional capacity is supposed to generate an increased income stream one period after it is acquired. The probabilities used to calculate the expected values were the probabilities of the different values of stock (S_i) occurring, given that the stock has been exploited in an optimal way for a long time.

In order that the said marginal increment in fleet capacity be worth while, the cost of the increase in capacity must be less than or equal to the increase in the present value of net income. Setting (4.29) equal to (4.30) makes it possible to calculate the critical investment cost per unit, c_k^*; that is, the cost per unit of capacity which implies zero profitability of a marginal increase in capacity:

$$c_k^* = (1 + r)^{-1}(1 + \delta/r)^{-1}[EV(\bar{Q} + \Delta Q) - EV(\bar{Q})]/\Delta Q \quad (4.31)$$

Figure 4.13 shows how the critical investment cost in kroner per kg of fish varies with the total fishing capacity measured in thousand tonnes of fish. As total capacity increases, a lower investment cost per unit of capacity is needed to make a marginal increase in capacity profitable. The actual investment cost depends on how much money needs to be invested to increase the fishing capacity by one unit, which again depends on the cost of the durable equipment needed in the fishery and the productivity of this equipment. In the example used above (Section 4.3.1), an investment of 50 million kroner was needed to increase fishing capacity by 25 000 t, which translates into a marginal investment cost of 2 kroner per kg of fish.

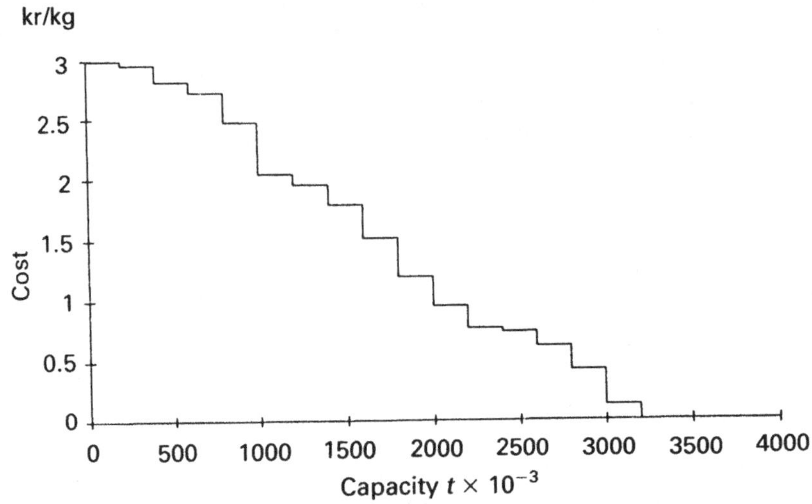

Fig. 4.13 Barents Sea capelin. Critical capacity cost.

The optimum capacity is located where the actual cost per unit of capacity is equal to the critical cost per unit. It is clear from Fig. 4.13 that the optimum capacity depends on the cost of capacity. This is in stark contrast with the deterministic case. As discussed in Chapter 3, the optimum fish stock and fishing effort in the case where the catch per unit of effort is constant and independent of the stock is determined by the discount rate and the marginal growth rate of the stock (the condition $r = G'(S)$). Here the optimum capacity depends on the cost of an additional capacity unit, and the optimum utilization of capacity depends on the state of the stock and the impact of current fishing on expected recruitment.

According to Fig. 4.13, the optimum capacity ranges from zero to 3.8 million tonnes, depending on the cost of a unit of capacity. If the cost per unit of capacity is greater than 3 (kroner per kg), the fishery is unprofitable. As the cost per unit of capacity decreases the optimum capacity increases, and at zero cost it would be worth while to invest in a capacity which can take the maximum catch of 3.8 million tonnes (this has been set arbitrarily in the model, as recruitments exceeding 3.9 million tonnes are very rare, but in theory it could be justified to invest in a capacity that can take the absolute maximum recruitment less the necessary spawning stock if the cost of capacity is zero).

This exercise with the stochastic recruitment function casts doubt on the validity of any advice on fishing capacity obtained from deterministic models which are meant to reflect average conditions in a stochastic environment. What the analysis here has told us is that the optimum fishing capacity depends on the degree of variability in fish stocks, besides the cost of fishing capacity.

The stochastic recruitment curve and random recruitment compared

The reason why the optimum capacity depends on the cost of fishing capacity in the model with the stochastic recruitment curve, contrary to the deterministic recruitment model, is basically the same as in the model with purely random recruitment. Here the total allowable catch is partly random; the recruitment resulting from any given spawning stock is uncertain but has a given probability distribution. If the cost of capacity is low enough, it is worth while to invest in a fleet which will often be partly idle but which can take advantage of large recruitments should they occur. The basic difference between the stochastic recruitment curve and the purely random recruitment is that with the recruitment curve the present utilization of fishing capacity has an impact on recruitment in the future, which must be taken into account.

According to Fig. 4.13, a capacity cost of 2 kroner per kg of fish, which corresponds to the assumed cost of fishing vessels, implies an optimum capacity of 1.2 million tonnes. Changing the cost of capacity to 1.8 would increase the optimum capacity to 1.6 million tonnes. When making comparisons with the purely random case above, we must keep in mind that we are dealing here with the entire capelin fishery, of which Norway takes 60 per cent, while the total allowable catch we considered in the purely random case was the Norwegian share. Hence the comparable optimum capacity here would be 720 000 t with a capacity cost of 2, and 960 000 t with a capacity cost of 1.8. This is less than the optimum capacity in the purely random case. It is not permissible, however, to draw any far-reaching conclusions on the basis of this comparison, since these two cases are in part founded on different sets of data and the parameters of the recruitment function are not known with a high degree of certainty.

In conclusion, let us again underline the main difference between the purely random case and the model with the stochastic recruitment function. In the purely random case the total allowable catch is $S - S_{min}$, the stock in excess of the minimum spawning stock to be left for precautionary reasons. With the recruitment curve the size of the stock to be left for breeding depends on the probability distribution of recruitments and the fishing capacity of the fleet. A large spawning stock is desirable in order to increase the probability of a large recruitment, but the usefulness of a large recruitment depends on the fleet capacity in place. If fishing capacity is cheap and the optimal fleet is large enough to take most of a large recruited stock, the fleet will not be used at all unless the spawning stock is large enough to ensure that a large recruitment is likely to occur. If on the other hand the fishing capacity is expensive and the optimum fleet is small, it will more often be fully used, as it does not make much sense to leave a large spawning stock behind in order to increase the probability of a large recruitment that cannot be utilized. A large spawning stock will then be the result of an insufficient capacity to harvest a large stock.

Chapter 5
Methods of Fisheries Management

As explained in Chapter 2, competitive exploitation of common property fish stocks is wasteful because each fisherman or company director does not take into account how his fishing effort affects the size of fish stocks and thereby their growth and the unit cost of fish. There are, in principle, two ways of dealing with this:

(1) Establishing property rights to fish stocks. Under a system of property rights, no one would have the right to fish from a stock except the owner and those who are explicitly authorized by the owner to do so. The owner could be a person, a company, or a collective of individuals. Clearly it would be in the owner's best interest to take into account how fishing affects the growth of the stock and the unit cost of fishing.
(2) Control of the fish stocks by the public authorities. Clearly the access to the fish stocks must be limited in some way in order to avoid the overexploitation resulting from open access. In this type of solution the fish stocks would remain common property in the sense that no one could claim exclusive rights to them unless being granted such rights by the public authorities. Note that the broader the collective of individuals who would be endowed with ownership to the resources in Solution (1), the less the difference between the two principal solutions.

5.1 Property rights to fish stocks

The property rights solution has clear advantages in theory. The owner of a fish stock has a strong incentive to limit fishing to whatever level maximizes profit, whether he cashes in the profit by fishing himself or by selling or leasing to others the right to do so. The profit-maximizing fishing effort will also be the socially optimal effort, provided the prices on the basis of which the profit is being maximized satisfy the following conditions:

(1) The owner of the fish stock must not be able to influence the price of fish or the price of inputs but must take them as being given by the market.

(2) The prices of fish and fishing inputs must reflect the social opportunity cost of fish and inputs.
(3) The discount rate used by the sole owner must be equal to the social discount rate.

Conditions (1) and (2) were explained in Chapter 1. As to the third, there exists an enormous literature dealing with the question what is the appropriate social rate of discount. This is not the place to review that discussion. One plausible school of thought is that the social rate of discount should be equal to the rate of return on marginal investment projects in the private sector. Applying this discount rate would ensure that the same criterion of profitability will be used for public investment as for private investment. There is some presumption that the interest rate charged by financial institutions will reflect this, as private investment projects financed by such institutions must yield a return that covers this interest cost.

There are reasons why the property rights solution has seldom if ever been tried and is not likely to be tried on an extensive scale. Most fish migrate widely, and the possibility of 'fencing them in' and effectively claiming ownership is small. For fish that straddle international boundaries the property rights solution would run into difficulties, as all private ownership depends ultimately on the legal apparatus and enforcement of a sovereign state. For fish that stay within the economic zone of one single state, the property rights solution is likely to be impractical and undesirable except for the most stationary stocks. For migratory stocks the property rights would have to be defined over a wide area in order to provide an effective control over each stock. This might give the property rights holder an undesirable degree of control over the market for fish or fishing inputs.

There are two ways in which property rights to fish stocks could be implemented. One is to make it illegal for anyone but the owner to land fish from a certain stock. This requires that it be possible to identify from which stock the fish originates, either by genetic characteristics or by knowing where it was taken. This is often possible. The other way is by territorial rights. To confer full control over a stock, the territorial right would have to apply over the entire area which a stock inhabits. Since fish stocks vary greatly in their migratory habits, an array of territorial rights differentiated according to stocks would have to emerge in order to ensure full control over all stocks.

It is possible, however, that territorial use rights in fisheries (TURFS), defined over some territory without any reference to the migratory habits of fish, would facilitate an efficient use of fish stocks. Even if such rights would not give any single holder of a territorial right full control over any single stock, they would provide an incentive to conclude mutually beneficial agreements on how to harvest the stocks. Collectively the holders of territorial rights would be interested in maximizing the aggregate profit of the fishery. Territorial rights would provide an incentive to do so, as they would limit the number of parties with access rights.

An agreement reached among the holders of territorial rights could thus not be challenged by a third party with no such rights. Unfortunately, however, the parties to a negotiated solution of this kind would also have an incentive to circumvent an agreement once it had been made and to catch as much fish as possible without being found out. Attempts at setting up cartels in various branches of industry or raw material production have typically failed for this reason.

The territorial use right arrangement has close parallels in the petroleum industry. Like fish, petroleum is a 'migratory' resource. Petroleum is extracted by piercing a hole in a natural underground reservoir where petroleum is trapped. The pressure in the reservoir forces the petroleum to migrate towards the hole. Drilling another hole would attract some of the remaining petroleum, and the owner of the second hole would be able to extract some oil that otherwise would have gone to the owner of the first hole. Since the right to drill holes into an underground oil reservoir is determined by the ownership of land above it, the analogy with territorial use rights in fishing is close.

How have territorial use rights worked in the petroleum industry? In certain places and at certain times (Eastern Texas in the 1930s) the situation in the petroleum industry has been similar to competitive exploitation of fish stocks. Competitive drilling of oil wells has increased extraction costs unnecessarily and reduced the recoverable amount of oil. The ownership of land above oil reservoirs in Eastern Texas was probably too fragmented for a negotiated solution to the extraction problem to emerge. By contrast, the oil companies holding exploitation rights to large blocks in the North Sea negotiate the sharing of oil fields that straddle the boundaries of these blocks, so discouraging overproduction by any single company. There is thus a presumption that a bargaining solution is more likely to work the fewer parties are involved.

In Japan fishermen's cooperatives have been given exclusive rights to inshore resources. These rights come close to amounting to ownership of the resources. The fishing rights appear successful in promoting efficient utilization of the resources. The fishermen's cooperatives regulate their members' access to the resources (others are excluded by the rights) and engage in stock enhancement such as the building of fish shelters and artificial reefs and the operation of hatcheries. There would be little incentive to engage in such activities without an arrangement which ensures that those who plant have the right to harvest. (On the Japanese fishing rights, see Asada *et al.* (1983), Ruddle (1984), Yamamoto & Shot (1992).

Even if property rights to fish stocks seem less than an obvious solution to the efficiency problem in fisheries, there are other types of property rights (rights to harvest) that appear easier to apply. These will be discussed below together with other methods of fishery regulation.

5.2 Why government regulation often fails

Given that private ownership of fish stocks is either impossible or impractical, we are left with government regulation as a means to achieve efficient use of these resources. This is certainly a type of solution that has been tried, in many if not most cases with meagre results. A discussion of government regulations of fisheries, therefore, appropriately begins by explaining why they have failed and what the remedies could possibly be.

Why have government regulations of fisheries failed? Most regulations of fisheries have been initiated when free access has caused severe and obvious depletion of the resource base, leading to low incomes or unemployment in the fishing industry. At this point there has emerged at least a degree of consensus that the resources must be safeguarded and replenished, in order to raise incomes and maintain employment in the industry. Fisheries biologists have been called upon to give advice on how to manage the resources to attain this goal. The regulations imposed to do so have typically aimed at limiting the catch of fish in order to protect the fish stock, such as closing the fishery for a certain period, prohibiting the use of certain kinds of fishing gear, etc.

These measures usually accomplish what they are primarily meant to accomplish, namely the protection of the resources. The side effect usually is to raise the cost in the fishery until all the fishing rent has been dissipated. As long as free access is preserved, revenues in excess of costs encourage further investment in the fishing industry. Even if the total amount to be caught is fixed, it still makes sense for the individual fisherman or fishing firm to invest in a bigger or better-equipped boat in order to obtain a larger share of the catch, as long as revenues exceed costs. Instead of free access competition leading to a depletion of the fish stock, it now leads to excessive investment in a self-defeating attempt to acquire a greater share of a given catch. In terms of overall economic efficiency and maximization of rents, nothing will have been achieved.

An example of this kind of remedy is the often proposed mesh size regulation in fisheries that exploit multi-year class stocks. In the absence of regulation of any kind, fishermen have an incentive to use nets that catch the fish at an early age, before they have attained the age of maximum year class weight (see the discussion of the Beverton-Holt model in Chapter 3). To rectify this it is frequently suggested that the minimum mesh size be increased, in order to utilize better the growth potential of the fish. The short-term effect of this would be to decrease catches, but after the young year classes that are being spared have reached fishable age the catch would increase for a given level of effort. If, however, nothing else is done, the fishery would in time attain a new equilibrium where revenues and costs match once more. As soon as the fish yield recovers and revenues start to exceed costs, an incentive arises either for some to increase their effort or for new firms or fishermen to enter the fishery.

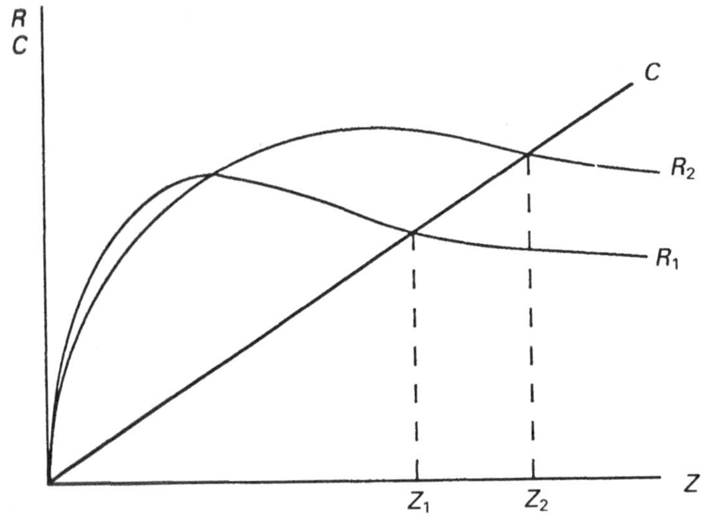

Fig. 5.1 Sustainable revenue curves and bioeconomic equilibrium for two different mesh sizes: (1) small; (2) large.

The argument is illustrated in Fig. 5.1. The two revenue curves refer to sustainable revenue for two different mesh sizes, R_1 for small meshes and R_2 for large meshes (curves of this kind can be derived from the Beverton-Holt model discussed in Chapter 3 by changing the selectivity parameter). Initially we are in a free-access equilibrium with small meshes with effort at Z_1. Then the minimum mesh size is increased. Initially the revenue would fall and effort might in fact decrease, but over time the revenue curve R_2 would be approached, profits would emerge, and fishing effort would start to expand. A new equilibrium would be reached at effort Z_2, where cost and revenue would be equal again. The only possible long-term benefit would be an increase in consumer surplus, as consumers would be able to buy more fish at a lower price. If the fish is exported, this gain would benefit foreign consumers.

5.2.1 *Incentive problems*

Why is it that fishery regulations so seldom aim at maximizing rents in the fishery and achieving economic efficiency? This is a little surprising, because governments do have some incentive to do so. Governments need sources of income to finance the provision of such goods and services as the market mechanism cannot be relied upon to provide in sufficient quantities. There is a case for tapping first and foremost those sources that interfere minimally with economic efficiency, unless this has clearly inequitable consequences. The potential conflict between efficiency and fairness may be illustrated as follows. Some goods (food, for example) are basic necessities. Putting a tax on such goods is likely to be more efficient than taxing luxury items, since people will buy basic necessities as long as they can afford it and will not reduce their work effort because they think basic necessities are not worth while to obtain at a high price. The unfairness of such a tax is obvious, given that people's incomes are different.

Taxes such as an income tax, a tax on profits (profit is used here in the sense of capital income and thus includes contribution to the covering of fixed capital cost in addition to pure profit), and a tax on sales or value added cause inefficiencies; income taxes discourage work effort, taxes on profits discourage investment, while a tax on value added is an indirect tax on the use of labour and capital. Taxes on the rent of natural resource extraction do not cause such distortions, except that in practice it is often difficult to design rules such that only the rent is being taxed and not income or profit. Partly for this reason the extraction of natural resources such as minerals (oil in particular) is heavily taxed in most countries. This raises the question 'why not tax fish?' Indeed, in some countries there is now an attempt to do so, particularly in New Zealand.

Besides being interested in generating a resource rent for use as a tax base, a government might be interested in an efficient fishing industry because it is interested in economic efficiency in general. After all, maximum material well-being requires maximum economic efficiency, even if some of that material well-being is provided by publicly produced goods and services financed by taxes. Nevertheless, although a high material standard of living is an important if not the only objective for governments and political decision makers in general, it does not follow automatically that public authorities will in fact try to manage fisheries for maximum rent. A possible reason is that objectives other than economic efficiency are being sought deliberately, despite their cost in terms of less economic efficiency. The legitimacy of such objectives is, however, often questionable. More often than not they appear to serve as vicarious arguments for serving special interest groups at the expense of the general public.

The reason why special interest groups often prevail at the expense of the general public is the asymmetric distribution of costs and 'gains' from economic inefficiency. The costs of economic inefficiency, such as unnecessarily high prices and a lesser economic growth than need be, are distributed among people in general. The 'gains' from inefficiency, such as higher incomes in an industry that is sheltered from international competition, by contrast benefit small, well defined, and often well organized groups. Hence, even if the gains from improving efficiency, such as by increasing competition in a particular market, outweigh the 'gains' from not doing anything of the sort, the 'gains' from the latter could appear more important and worth fighting for.

This asymmetry can make it worth while for governments to seek popular support through promoting special interests by increasing inefficiency rather than promoting the general interest by increasing efficiency. This is probably a fairly general tendency, up to a point where the inefficiency becomes too blatant. All governments need some popular support, even those who by no stretch of the imagination could be called democratic. The support of a particular group, such as those who earn their living in the fishing industry, will most likely be forthcoming if the special interests of that particular group are being promoted. Such special interests can easily be, and often are, contrary to the interests of the general public. The fishing industry will benefit from high prices of fish brought about by restrictions on imports of fish (if the country in question is not self-sufficient in fish), from support of fishermen's incomes financed by general tax revenue, or by not having to pay access fees or abide by other restrictions that limit overexploitation of fish stocks. All of these are important matters for fishermen and people employed in the processing industry, but may account for little for each individual taxpayer or consumer who must pay for it.

The conclusion this argument leads to is rather depressing. Even benevolent and well-informed governments will be tempted to pursue policies that lead to lesser material well being than need be, without any obvious non-material benefits in return, because of the dispersion of the benefits of greater efficiency and the concentration of gains accruing to special interest groups from regulated markets and inefficient use of resources.

This neglect of the economic efficiency of the fishing industry is reinforced by the fact that it is often a small part of a nation's economy, but locally concentrated. Political representatives from areas in which the fishing industry is concentrated come under pressure from the industry to support it financially if there are not enough fish around. The industry typically resists structural changes which are likely to be painful in the short term, even if they promise a greater future gain. This is understandable, given that the future is always uncertain. Political representatives are likely to attach a greater value to their chances of getting re-elected than to issues such as economic efficiency and government finances. The costs of bailing out the fishing industry is spread over a large population and of limited concern to each individual.

So much for the incentives faced by politicians, but what about the civil servants whose task it is to manage the fishery? In theory civil servants are subordinate executors of decisions made at the political level, but in practice they do influence such decisions to some degree, partly because they prepare the technical material on the basis of which the political authorities take their decisions, and partly through the nitty-gritty details that for practical reasons are left to civil servants to decide. What, then, are the incentives faced by civil servants? Can they be expected to promote efficiency in the fishing industry?

Civil servants can hardly be expected to try to achieve maximum profit in an industry they are supposed to regulate unless they are given a clear mandate to do so and their salaries and careers made contingent upon how successful they are. Without such incentives, civil servants are more likely to see themselves as referees in a sportsmanlike game for hunting fish, trying to achieve equal and equitable treatment of the players in the game. Indeed their careers are likely to be determined on the basis of how well they accomplish this and how aptly they follow established and formal rules of procedure. Furthermore there are often informal and personal ties between the public regulators and the industry being regulated. Public administrators often become employed by the industry they used to regulate. As lobbyists they know from their previous experience which doors to knock on in order to get the most favourable regulation.

5.2.2 *Efficient and fair regulations*

Is public regulation of the fishing industry doomed to failure? Not necessarily. The challenge lies in devising a regulatory mechanism that emulates a perfect market and is as independent as possible of political decision-making. One step in this direction would be to set up a public management authority with an explicit mandate to maximize the rents in the industry and to reward its managers according to how well or badly they perform. An agency like that would most likely see an advantage in relying on efficiency-promoting mechanisms like individual transferable quotas, which will be discussed further below. In fact management bodies might be largely or wholly redundant, as systems with inherent tendencies to promote efficiency could do the job on their own, with minimal interference by public authorities.

Before we discuss the management methods that seem most likely to promote economic efficiency, let us consider briefly how such methods, or any other change in management which would improve efficiency, might be implemented with the consent of the industry. Suppose we have a fish stock which is declining rapidly. It is decided to try to reverse this and to rebuild the stock to a more appropriate level. To achieve this, fishermen will have to put up with at least a temporary restraint on their activity. This they are often reluctant to do, even if they are promised a future gain in the form of a more secure employment and higher incomes. What they will most likely fight against is any forced reduction in the number of boats and fishermen. Since moving towards increased efficiency from a position of free access and no rent implies some reduction in fishing effort, this will be difficult to attain with the consent of the industry, unless those who leave the industry get a compensation making them no worse off than they would be if they had continued fishing.

Irrespective of whether fishermen deserve to be compensated for the losses they would incur for rebuilding a fish stock (after all, they depleted it in the first place), it is important to note that such compensation is possible, and the very fact that it is possible is itself the reason that a replenishment of the stock is indeed worth while. Increased economic efficiency means that a surplus will emerge, in this particular case an excess of revenues over costs. Those who participate in the fishery at the time when the replenishment of the stock starts can share in this gain, so that no one need come out worse than if the transition to the more efficient situation was never made. Since the gains take time to emerge, the government or the industry itself would have to advance funds to compensate potential losers. A practical example of this mechanism is the introduction of individual transferable quotas in the inshore fishery in New Zealand (see Clark *et al*. 1989). About 20 per cent of the fishing quotas allocated on the basis of catches in the immediately preceding years were bought back from the industry in order to rebuild the fish stocks by reducing the total catch.

5.3 Characteristics of efficient management

Management systems with built-in tendencies to promote maximum profitability of the fishing industry can be characterized by two phrases, compatibility of incentives and efficient use of information. Compatibility of incentives means that each decision-maker in the industry serves the public interest together with his or her own. The importance of such arrangements can hardly be overstated. The motive to serve one's own interests before anything else is a strong one in human beings. The ability to bend administrative orders, manipulate information, etc., while having in principle to obey instructions from above, is often great. Hence administrative orders that run counter to individuals' self interests will often

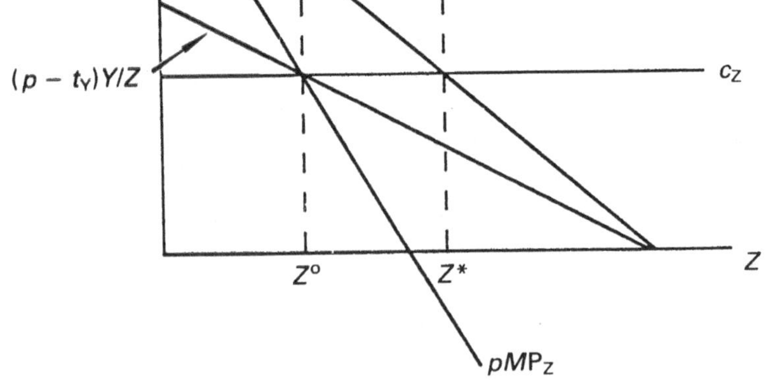

Fig. 5.2 Achievement of optimal effort (Z°) by a tax on effort or catch.

accomplish little, compared with systems in which the interest of the individual and that of the company or society are served by the same type of action.

This brings us to the other phrase, efficient use of information. Central planning requires a lot of information to be gathered and processed. Such gathering is costly, and the information is often distorted, for reasons already discussed. Systems that build on incentive-compatible use of information at the level where it is collected as a matter of necessity are more cost-efficient than systems requiring extensive gathering and processing of information. To give an example, the individual skipper will usually know best where, when and how to fish in order to get the most profitable catch. If his rewards for fishing are not sufficiently related to the success of his fishing trip, he may find it more advantageous to go somewhere else where less effort is required, and he may have his ways of convincing his superiors that he did the right thing.

What are the principal ways in which governments may control the fishing industry in an incentive-compatible and information-efficient way? In the discussion to follow we shall distinguish between three of these:

(1) indirect controls through a tax on fishing effort or the landings of fish;
(2) control of fishing capacity and fishing effort (use of available capacity);
(3) control of the catch.

5.4 Indirect control through a tax on landings or effort

Fishing effort tends to be excessive because the fishing industry pays no price for access to the fish even if they are scarce. The scarcity of fish stocks as a factor of production is due to two effects. First, the growth potential of fish stocks will be diminished if they are reduced below a certain level (the MSY level). Second, the catch per unit of effort is likely to be less the smaller the stock from which it is taken, implying that the cost per unit of fish landed will be higher the smaller the stock is. Putting a tax on the landings of fish or the use of fishing effort is a way to put a price on extracting fish from the stock. An appropriate tax will in theory keep the fishery in a long-term optimum. To see how this works, consider Fig. 5.2.

The figure shows the value of the sustainable catch per unit of effort (pY/Z) and the (undiscounted) value of the marginal sustainable catch of effort (pMP_z) in the case when the catch per unit of effort is proportional to the size of the stock, at a constant price of fish. As explained in Chapter 2, the long-term optimum will be at the point of intersection between the value of the marginal catch of effort and the marginal cost of effort (c_z), as shown in the figure by the point Z^o. As also explained in Chapter 2, the open-access fishery will be in a long-term equilibrium where the marginal cost of effort is equal to the value of the catch per unit of effort, at point Z^* in the figure.

As can be visualized from Fig. 5.2, it would clearly be possible to make the economic equilibrium point coincide with the optimum by manipulating the marginal cost of effort or the value of the catch per unit of effort. If the cost of effort were raised by a tax t_z per unit of effort, as shown in Fig. 5.2, the economic equilibrium point would coincide with the economic optimum. The free-access fishery would then attain the economic optimum without any further interference simply by having to adjust to the effort tax. The effort tax can thus be regarded as the adequate price to pay for access to the fish stock. Applying this price would make the free-access solution efficient, and the deficiencies of the market mechanism under free access would be corrected for.

This solution could also be attained by a tax on the landings of fish. The net price to the industry would be $p - t_y$, t_y being the tax per unit of fish. The line $(p - t_y)Y/Z$ shows the value of the catch per unit of effort at the net price received by the industry. By a suitable manipulation of the tax rate, it would be possible to make this line intersect the cost per unit of effort line right above the point Z^o, so that the free-access equilibrium would coincide with the economic optimum.

Neat as these solutions are in theory, they are unlikely to work well in practice. The main reason is that the abundance of fish stocks fluctuates very substantially over time, as discussed in Chapter 4. Whether or not the appropriate response to these fluctuations is to maintain a stable catch or to let it fluctuate with the stock, the tax rate would in all probability have to be varied from one year to another, in order to get precisely the amount of effort and catch that would be appropriate. These variations would have to be determined on the basis of reliable knowledge of how the industry would respond in the short term to variations in the tax rate. Such knowledge is not likely to be available and is in any case costly to gather. To give an example, it is quite possible that the short-term response to a change in the tax rate will be the opposite of the long-term response depicted in Fig. 5.2. Fishing firms must in the short term pay interest and amortization on the loans they have acquired, and if it becomes more difficult to do so because the price of fish has come down (because of a higher tax), they might choose to increase their effort in order to make ends meet. This is particularly likely to happen in one-man firms where the firm's productive effort is the result of individual choice between effort and leisure. This does not in any way contradict the result above that a higher tax will lead to less fishing effort in the long term. A higher tax will make investment in fishing vessels less profitable, so when the time comes to make a decision whether or not to replace worn out vessels, fewer people will find it in their interest to do so if the tax on fishing has been raised.

For the other case that we have used extensively for purposes of illustration (a constant catch per unit of effort no matter how large or small the stock is) the tax solution would be even less adequate. Disregarding fluctuations in the stock, it is possible to imagine that the optimal tax could be found by trial and error in the case shown in Fig. 5.2. If the tax were set too low, one would observe too many

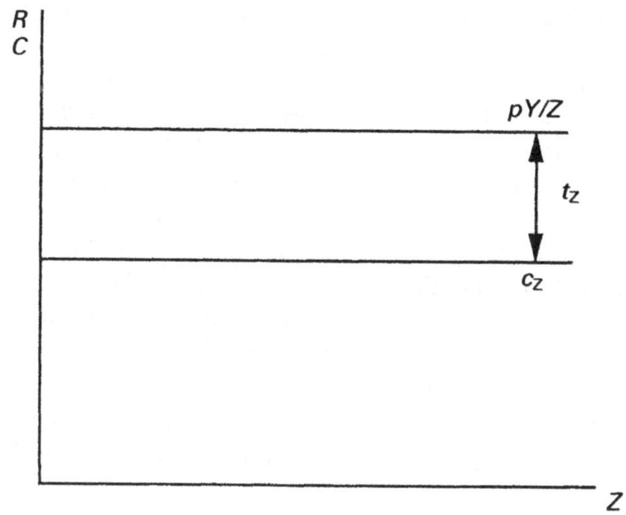

Fig. 5.3 The impossibility of using a tax to achieve optimum effort when the catch per unit of effort is constant.

boats in the fishery. That could be corrected by raising the tax, and gradually the number of boats would approach the optimum. When the catch per unit of effort is constant, as shown in Fig. 5.3, the optimum solution is extremely unstable. The optimum tax fills the gap as it were between the cost per unit of effort and the value of the catch per unit of effort, but the slightest deviation from the optimum level implies that the fishery is either unprofitable or profitable in the long run at all levels of effort. If profitable, the fishing fleet would expand until the stock had been totally depleted. Finding the optimal tax by trial and error could be a costly exercise, leading to a total depletion of the stock before the correct tax level had been found. For stocks characterized by this production relationship, as some pelagic stocks appear to come close to being, the tax solution is not likely to be adequate.

There is, nevertheless, one desirable property of the optimum tax solution worthy of note. The industry would be left to conduct its own affairs on the basis of the information it normally gathers as an ordinary business routine. There would be no gathering and processing of information about the fishing industry by the managing authority in charge of the fishery, except to try to find out as precisely as possible how the industry would react to a tax and to hit the correct tax rate as quickly as possible. Neither would there be any commands to be handed down to the industry. Thus the tax solution uses information efficiently. It is also incentive-compatible. Decision-makers in the fishing industry would be acting in the public interest (maximizing the resource rent) at the same time as they are maximizing their profits, provided that they have no control over the market prices of fish and of fishing inputs. These are properties to be looked for in alternative and more practical solutions.

Although the tax solution is not likely to be practical in most cases, a tax on landings or effort would be an appropriate device to syphon off some of the resource rent for the benefit of the public. This kind of tax has the desirable property of encouraging efficiency in the fishing industry, provided that it is not excessive, even if it is not sufficient for achieving efficiency. A tax on fish is likely to be preferable to a tax on effort, because of the multidimensionality of effort. Effort is composed of many different factors; capital, labour, and various inputs. All of these must be taxed proportionately in order not to distort the cost-minimizing combination of inputs. If, for example, the use of capital in the fishery is taxed but not the use of labour, the industry will prefer boat designs that use less capital and more labour than otherwise. This is undesirable if the cost of labour and capital in the absence of a tax reflects the true social opportunity cost of these factors.

5.5 Control of fishing capacity and effort

Given that the tax solution is not the way to achieve optimality, we must look at more direct ways of control. One such way is to control the fleet capacity and its use by fishing licences. Relying exclusively on this method requires two types of licences for each vessel, a long-term licence without which a vessel owner has no right to participate in a fishery, and a short-term licence giving the owner permission to do so for a given period of time. The long-term licences would control the aggregate fleet capacity. Because of fluctuations in fish stocks and total allowable catches, it will in many cases be necessary to control the use of the available fishing capacity, as discussed in Chapter 4. This would be the purpose of the short-term licences. Because the short-term relationship between fishing effort (use of fishing capacity) and the catch is likely to be uncertain, it is likely that fluctuations in the allowable catch will be better dealt with by catch quotas than by limitations in the use of fishing vessels. A possible exception are fisheries in which it is particularly difficult to predict the size of fish stocks and hence to set an allowable catch before a fishing season starts. In such cases it could be better to monitor the catch per unit of effort as the fishing proceeds, provided that it is a good indicator of the abundance of fish, and stop fishing when the catch per unit of effort has fallen below a certain critical limit. Squid may be an example of stocks for which this is an appropriate method.

The control of fleet capacity may appear straightforward and preferable to a control of the catch, since excessive fishing capacity is the fundamental problem to be dealt with by fisheries management. In practice this is not so. The reason is again the multi-dimensionality of fishing effort and fleet capacity. For practical reasons it will be possible to control only a few important dimensions of fleet capacity. Controlling the number of vessels alone will not be enough; fishing vessels vary in size and other characteristics important for their fishing capacity. A permit to own and operate a vessel without any further specification would give the licence holder an incentive to acquire a large and well-outfitted vessel, and the fleet capacity would rapidly get out of control if the licensing leads to improved profitability. To deal with this the licences might specify the cargo capacity of a vessel, as is done in the Norwegian purse-seine fishery. It might even be necessary to specify other characteristics as well, as it is usually possible to increase the fishing power of a vessel by acquiring more powerful engines, fish finding equipment, etc. Fishing capacity is likely to expand along uncontrolled and uncontrollable dimensions, and since for practical reasons effort and fleet control can hardly be all-inclusive, it is likely to distort the cost-minimizing combination of inputs. More serious, perhaps, is that this type of control may 'freeze' the technology and prevent the development of vessels with a new and better design and equipment which happen not to be compatible with the regulation. Instead, the ingenuity of naval architects is likely to be directed toward developing vessel designs that circumvent the regulation and pack a much-increased fishing power into a hull that formally satisfies the requirements. In Norway the term 'paragraph trawler' is used of vessels that measure 299 GRT and satisfy the required size limit of 300 GRT while having the fishing power of a vessel that would ordinarily measure more than that.

5.6 Control of the catch

Catch control is a widely used method of regulation. Setting a limit to the total catch that can be taken, however, accomplishes nothing in economic terms unless it is supplemented by additional measures. If free access is allowed to continue while an upper limit is imposed on the total catch, there will be intensified competition for the largest possible share of it. The fish stock will be preserved at a healthy level, provided that the catch limit is correctly set and efficiently enforced, but the improved profitability generated by rebuilding fish stocks will only fuel wasteful competition in the industry. This competition will, as otherwise under free access, proceed to the point where the cost equals the revenue. One of the most flagrant examples of a regulation of this kind is the American Pacific halibut fishery. Because of ever-expanding fishing fleets, the fishing season has shrunk from several months to a few days per year. Much of the catch is processed into products of lower quality than need be if it were spread over a longer period of time, and the attempt to catch as much as possible in a short period of time adversely affects the quality of the fish landed (see, for example, Richards & Gorham 1986). This is just as wasteful as the depletion of fish stocks.

The obvious way of avoiding this self-defeating competition is to divide the permitted catch between the boats that participate in the fishery. Everyone will then know how much he can take, and the incentive to get as large a share of the total catch as possible by investing in more, bigger and more powerful boats will be eliminated. Furthermore there is an incentive to make the greatest possible profit from the permitted catch by keeping down costs and producing as valuable product as possible.

The efficiency gains can be made greater still by making catch quotas transferable. Transferability gives less efficient holders of quotas an incentive to sell them to others who are able to use them more efficiently. Transferability will in the short run lead to a better utilization of a given fleet. When the total allowable catch is small and each vessel gets a small quota, it is likely that all the boats available will not be needed. If the most efficient operators are permitted to buy quotas from the less efficient ones at a price which is mutually beneficial to both, redundant boats will be laid up while the rest can be used at full capacity. Transferability in the long run means that those who leave the fishery can sell their quotas to others who want to enter the fishery or who want to acquire a larger and more economical boat. Transferability will thus lead to a better matching of fishing capacity and the availability of fish, both in the short and the long term.

Some countries have introduced individual transferable quotas in a more or less 'pure' form in recent years, most notably New Zealand, Australia, and Iceland. Transferable quotas satisfy the conditions for an efficient management that were discussed above. They are incentive-compatible, as fishermen and fishing firms strive to make the most from their quotas, which contributes to maximizing the rent in the fishery. They are efficient in the use and processing of information, as all decisions, except how large the total allowable catch should be, are taken by the firms themselves.

5.7 Transferable quotas

Since transferable quotas appear particularly promising as an incentive-compatible and cost-efficient system of management, we shall now discuss them in some detail. A question that is natural to ask first is how is it possible to provide fishing firms with long-term quotas, given that fish stocks and the total allowable catch often fluctuate in an unpredictable manner? If fishing firms are given a right to a long-term fixed quota, possibly for an indefinite period of time, the public authorities would have to buy and sell quotas each year in order to keep the catch within the limit set by the total allowable catch. In lean years they would have to buy quotas from the industry in order to limit the catch, while in good years they could make up for this by selling excess quotas to the industry.

In order to ensure that the government does not lose money on these transactions, and perhaps makes a net gain, the total amount of quotas issued must be less than the maximum possible total allowable catch. Setting the appropriate limit to the total amount of quotas is not a trivial exercise, however. The total amount of quotas must not be set so high that the government will be buying quotas from the industry most of the time and losing money. If, on the other hand, the total amount of quotas is set at a 'low' level, the government will be selling excess quotas most of the time and making money, which is less of a problem. Setting the total amount of allocated quotas equal to the average total allowable catch does not necessarily accomplish the goal of ensuring a financial balance for these transactions. Which level of the total amount of quotas allocated accomplishes long-term financial balance will depend on how the price of quotas varies with the total allowable catch. The market price of fish is likely to vary inversely with the quantity caught, so if nothing else were involved the price of quotas traded would vary inversely with the total allowable catch. Then, if the government would set the total amount of quotas equal to the average total allowable catch, it would be buying dear and selling cheap and making a loss in the long term. In order to come out even or to make money the government would have to set the total amount of quotas below the average of the total allowable catch.

The relationship between the total allowable catch and the price of quotas could, however, be the opposite. When the total allowable catch is large, the stock will be plentiful and the catch per unit of effort is likely to be high. The cost per unit of landed fish would be low, and the difference between a constant market price and the cost per unit of fish caught would be great. Hence the price of quotas traded would be high, so if the government had set the total amount of quotas equal to the average total allowable catch, it would be buying cheap and selling dear, thus syphoning off some rent on the average. How the total amount of quotas should be set in order to ensure a long-term financial balance from the buying and selling of quotas is considered in Hannesson (1989).

5.7.1 *Quotas: fixed quantities or shares in the total allowable catch?*

As prices and costs change over time, the total quantity of quotas yielding a financial balance in the quota transactions undertaken by the government will change. This makes it a doubtful proposition to determine individual quotas as fixed quantities for the very long and possibly indefinite future, as it would imply a considerable financial risk for the public authorities. This can be easily avoided by determining individual quotas as shares in the total allowable catch (the Icelandic quotas, and since 1990 the quotas in New Zealand, are determined in this way). The individual quotas will then vary with the total allowable catch, and the government will be relieved of the need to buy and sell quotas in order to keep the catch within the necessary limits. Such share quotas nevertheless provide individual firms with incentives to invest optimally in relation to their quotas. On the basis of their shares and the available biological knowledge about fish stocks, fishermen will be able to predict as precisely as the available information allows what their quotas are likely to be in the future and invest accordingly.

One (and perhaps the major) difference between transferable quotas determined as shares of the total allowable catch versus fixed quantities is that the industry has to bear all the risk associated with variations in the total allowable catch in the former case, while with fixed quotas the industry is guaranteed a certain income. If the fishing firms are not allowed to catch the sum total of their fixed quotas, the government will have to buy a part or all of it. This does not, however, necessarily imply that the income in the industry will be more stable with fixed quotas. When the price of fishing quotas varies inversely with the amount of fish caught, the government will be buying quotas at a high price and selling them at a low price. The inverse relationship between price and quantity landed will tend to stabilize the revenue in the industry, compared with a situation in which the price is constant. Buying excess quotas from the industry at a high price may in that case increase the variations in its revenue (see Hannesson 1989).

5.7.2 *Are transferable quotas fair?*

As already discussed, transferable quotas are likely to lead to efficient utilization of fish stocks. But is this a fair method of management? We may distinguish between two aspects of this: (1) how catch quotas could be introduced in an equitable way, and (2) how quotas could function equitably once in place.

Above we have discussed a possible criterion for an equitable change in policy; that no one who will be affected should lose. This is an ideal unlikely to be fully attained, but that is not an argument against trying to approach it as closely as possible. An initial allocation of quotas on the basis of recent performance appears fair and has been practised in most places where transferable quotas have been introduced (e.g., New Zealand, Iceland). This defines a reference position reasonably close to what can be expected to prevail if no regulation were put in place. Other criteria are also possible. For instance, when catch quotas were introduced in the Australian bluefin tuna fishery, a formula based on previous catches and investment in fishing vessels was used, and the weighting of these factors was determined by the spokesmen for the fishing industry (see Robinson 1986). A free trade in quotas will ensure that those who want to leave the fishery will get a share of the benefit of a more efficient industry, by being able to sell their quotas to others who are better able to use them.

The main problem likely to arise in connection with an initial allocation based on recent performance is whether to allocate quotas only to the owners of fishing vessels or to include also those who work for a wage or a share in the catch. In large-scale fisheries where a large part of the vessel crew is hired labour and the owners may not be active fishermen, the fairness of allocating quotas only to the boat owners is most questionable. The large number of people who would be involved and the high turnover of labour that often characterizes such operations are arguments for restricting the allocation to the boat owners only. This is, however, a practical argument and not one of principle.

The fairness of transferable quotas once they are in place is closely tied to the question of who owns the fish stocks and who should be entitled to the rent they can provide. According to the Law of the Sea Convention, the coastal state has exclusive rights to fish resources within 200 miles from shore but is obliged to allow others to fish that part of the total allowable catch which it cannot take itself. This obligation seems, however, not to be very realistic, as the coastal state itself determines the total allowable catch and how much it can take. This would seem to imply a form of collective ownership of the fish resources of the people of the coastal state. Even if the coastal state presumably is as free to privatize these resources as it is with respect to other assets, such schemes are likely to be controversial, and the question will arise with what right a selected few who happened to be in the industry when the fishing quotas were introduced are to be favoured before everybody else.

Since a management system based on rights of access has clear advantages in terms of efficiency, it seems very much worth while to try to devise a method to make such rights compatible with public ownership and rights to resource rents. Such methods are not difficult to find. Other extractive industries, in particular the petroleum industry, can be used as models. The coastal state owns petroleum reservoirs underneath the seabed of the continental shelf but sells or leases extraction rights to private companies. The state gets a share of the petroleum rents by selling or leasing the rights to extract, and also by special taxes on petroleum.

Three principal ways of extracting fishing rents can be distinguished:

(1) the right to fish (quotas) can be sold once and for all to private entrepreneurs;
(2) the rights can be leased for a limited period of time;
(3) the rights can be allocated to private interests free of charge, with a special tax on fishing imposed in order to extract rents.

Each of these methods has its advantages. The once for all sale implies that the government transfers the ownership of the resources to private interests once for all. This is a drastic measure, and it is not likely to bring in more than a fraction of the present value of the resource rent. As the extraction of a renewable resource stretches into the indefinite future unless the resource is destroyed, the present value of fishing rents could be quite high, particularly at low rates of discount. Companies wishing to buy the rights to exploit such resources will, however, be reluctant to pay the full present value of the rents these rights confer, partly because these rents are uncertain and partly because they would not be able to finance such large outlays. A government that wants to extract most of the future rents would not be well advised to use this method, unless it were supplemented by a special tax on fishing. Such once for all transfer to the private sector has, however, the advantage that the system would be more firmly entrenched legally. It would be difficult for short-sighted politicians to tinker with the system, as it would amount to an abrogation of rights and be equivalent to confiscation of property.

If the government retains the formal rights to the fish resources, it can lease the rights to use them for a period at least as long as the economic lifetime of a fishing boat. Competitive selling of such leases of a finite duration, for example by auctioning, is more likely to bring a large share of the future rents into the public coffers than selling fishing rights once and for all. Depending on the terms of the lease and the stability of the political and judiciary institutions in the country, this could go a long way towards matching the system of permanent rights in terms of stability and predictability.

A special tax on fishing can be applied to the system with privately-owned quotas of an indefinite duration, just as it is applied to the extraction of minerals and oil on private lands. Such taxes can also be applied when the quotas are allocated to private individuals or firms for a shorter period of time, instead of, or as a supplement to, other charges. The disadvantage of a tax compared with competitive selling of fishing rights is its degree of arbitrariness. While a company would never be willing to pay more than its expected rent for a quota (that the expectation could be wrong is a separate issue), the tax authorities might be greedy enough to set taxes that would exceed the rents in the fishing industry. They would not, however, be able to do so for long, as the fishing fleet would not be renewed.

5.7.3 *Critical factors: control and enforcement*

The main drawback of quotas is the control needed to make them effective. Quotas provide incentives for misreporting, as has been demonstrated wherever quotas, transferable or not, have been put in place (Iceland, New Zealand, the European Community). Every fish not reported is a licence to catch the same quantity later in a legal manner. These problems are likely to be formidable to the point of being too costly to overcome in places where fish are landed from small craft with a minimum of technical devices and sold directly to consumers. In large-scale operations where elaborate landing facilities are required and where most of the fish is processed on shore, it is not too difficult to keep track of the fish and check different parts of the catching and processing system against one another. This method has been implemented in New Zealand and in Iceland. Problems are also likely to arise at sea. The system gives fishermen an incentive to 'high-grade' catches; that is, to throw away less valuable fish in order to accommodate more valuable fish within the limits of the quota. Besides being wasteful, such practices could have serious consequences for management, as the recorded catches would underestimate the quantity actually removed from the stock. This would weaken the factual base on which decisions about total allowable catches must be taken.

5.7.4 *Fleet and effort control reconsidered*

In case catch quotas are difficult to implement, boat licences might be a better alternative. Boat licences are much easier to monitor: the boats are there to be seen and counted, and they are not easily hidden. With the imperfections that always seem likely to be associated with a boat-licensing scheme, it certainly is easier to enforce. In general, it is important to note that the cost of enforcement is an essential element in evaluating a management scheme. A management system that appears efficient but is costly to enforce could be much inferior to a scheme giving rise to some economic distortions but which is less costly to enforce.

The arguments above about transferability and fairness of catch quotas also apply to effort controls through the licensing of boats. Boat licences can be made transferable, so that the licences end up being held by those who can use them most profitably. The licences can be introduced equitably by licensing those who are in the industry at the time, thus giving everybody a share in the potential rent of the fishery (such methods have been used virtually everywhere licences have been introduced; the Norwegian purse-seine fleet is one example). Like catch quotas, boat licences can be leased or sold for the purpose of getting a share of the resource rent.

References and Selected Bibliography

Andersen P. & Sutinen J.G. (1984) A survey of stochastic bioeconomics: methods and results. *Marine Resource Economics*, **1**, 117–36.

Anderson L.G. (1986) *The Economics of Fisheries Management*. Johns Hopkins University Press, Baltimore.

Asada Y., Hirasawa Y. & Nagasaki F. (1983) Fishery Management in Japan. FAO Fisheries Technical Paper, No. 238.

Berck P. (1979) Open access and extinction. *Econometrica*, **47**, 877–82.

Beverton R.J.H. & Holt S.J. (1957) On the dynamics of exploited fish populations. *Fishery Investigations*, Series II, Vol. 19. Her Majesty's Stationery Office, London.

Bjørndal T. (1987) Production economics and optimal stock size in a North Atlantic fishery. *Scandinavian Journal of Economics*, **89**, 145–64.

Bjørndal T. & Conrad J.M. (1987) The dynamics of an open access fishery. *Canadian Journal of Economics*, **20**, 75–85.

Butterworth D.S. (1981) The value of catch-statistics-based management techniques for heavily fished pelagic stocks with special reference to the recent decline of the Southwest African pilchard stock. In *Applied Operations Research in Fishing* (Ed. by K.B. Haley) pp. 441–64. NATO Conference Series II, Vol. 10. Plenum Press, London.

Charles A.T. & Munro G.R. (1985) Irreversible investment and optimal fisheries management: a stochastic analysis. *Marine Resource Economics*, **1**, 247–64.

Christy F.T. Jr. (1975) Property rights in the world ocean. *Natural Resources Journal*, **15**, 695–712.

Clark C.W. (1973) When should whaling resume? Department of Mathematics, University of British Columbia, working paper.

Clark C.W. (1976) *Mathematical Bioeconomics*. Wiley, New York.

Clark C.W. (1985) *Bioeconomic Modelling and Fisheries Management*. Wiley, New York.

Clark C.W., Clarke F.H. & Munro G.R. (1979) The optimal exploitation of renewable resource stocks: problems of irreversible investment. *Econometrica*, **47**, 25–49.

Clark I., Major P. & Mollett N. (1989) The development and implementation of New Zealand's ITQ management system. In *Rights Based Fishing* (Ed. by P. Neher, R. Arnason & N. Mollett), pp. 117–45. NATO ASI Series E, Vol. 169. Kluwer, Dordrecht.

Commonwealth of Australia (1989) *New Directions for Commonwealth Fisheries Management in the 1990s*. Department of Primary Industry and Energy, Canberra.

Conrad J.M. (1987) *Natural Resource Economics: Notes and Problems*. Cambridge University Press, Cambridge.

Conrad J.M. (1989) Bioeconomics and the bowhead whale. *Journal of Political Economy*, **97**, 974–91.

Copes P. (1972) Factor rents, sole ownership, and the optimum level of fisheries exploitation. *The Manchester School of Social and Economic Studies*, **40**, 145–63.

Crutchfield J.A. & Pontecorvo G. (1969) *The Pacific Salmon Fisheries: A Study in Irrational Conservation*. Johns Hopkins University Press, Baltimore.

Crutchfield J.A. & Zellner A. (1962) *Economic Aspects of the Pacific Halibut Fishery*. Fishery Industrial Research, Vol. 1, No. 1. U.S. Government Printing Office, Washington, DC.

Cunningham S., Dunn M.R. & Whitemarsh D. (1985) *Fisheries Economics: An Introduction*. Mansell Publishing, London.

Dorfman R. (1967) *Prices and Markets*. Prentice–Hall, Englewood Cliffs.

Dreyfus S.E. & Law A.M. (1977) *The Art and Theory of Dynamic Programming*. Academic Press, New York.

Flåten O. (1987) Sesongvarierende bestandstilgjengelighet og produktfunksjoner i Lofotfisket (Seasonal variations in the catchability coefficients and production functions in the Lofoten fishery; in Norwegian). Working paper No. 7005/1–77, FORUT, University of Tromsø.

Gordon H.S. (1954) The economic theory of a common property resource: the fishery. *Journal of Political Economy*, **62**, 124–42.

Hannesson R. (1975) Fishery dynamics: a North Atlantic cod fishery. *Canadian Journal of Economics*, **8**, 151–73.

Hannesson R. (1983) Bioeconomic production function in fisheries: theoretical and empirical analysis. *Canadian Journal of Fisheries and Aquatic Sciences*, **40**, 968–82.

Hannesson R. (1987) The effect of the discount rate on the optimal exploitation of renewable resources. *Marine Resource Economics*, **3**, 319–29.

Hannesson R. (1989) Catch quotas and the variability of allowable catch. In *Rights Based Fishing*. (Ed. by P. Neher, R. Arnason & N. Mollett), pp. 459–65. NATO ASI Series E, Vol. 169. Kluwer, Dordrecht.

Hannesson R. & Steinshamn S.I. (1991) How to set catch quotas: constant effort or constant catch? *Journal of Environmental Economics and Management*, **20**, 71–91.

Kirby M.J.L. (Chairman) (1982) *Navigating Troubled Waters. A New Policy for the Atlantic Fisheries*. Report of the Task Force on Atlantic Fisheries. Ministry of Supply and Services,Ottawa.

Larrañeta M.G. & Vazquez A. (1982) On a possible meaning of the polar motion in the ecology of the North Atlantic cod (*Gadus morhua*). International Council for the Exploration of the Sea, C.M. 1982/G:14.

Lindgren B.W. (1968) *Statistical Theory*. Macmillan, London.

Mollett, N. (Ed.) (1986) *Fishery Access Control Programs Worldwide: Proceedings of the Workshop on Management Options for the North Pacific Longline Fisheries*. Alaska Sea Grant, Report No. 86–4. University of Alaska, Fairbanks.

Munro G.R. (1982) Fisheries, extended jurisdiction and the economics of common property resources. *Canadian Journal of Economics*, **15**, 405–25.

Neher P.A. (1990) *Natural Resource Economics. Conservation and Exploitation*. Cambridge University Press, Cambridge.

Neher P., Arnason R. & Mollett N. (Eds) (1989) Rights Based Fishing. NATO ASI Series E, Vol. 169. Kluwer, Dordrecht.

Pearse P.H. (Commissioner) (1982) *Turning the Tide. A New Policy for Canada's Pacific Fisheries*. The Commission on Pacific Fisheries Policy, Final Report. Vancouver.

Richards H. & Gorham A. (1986) The Demise of the U.S. Halibut Fishery. In *Fishery Access Control Programs Worldwide*. (Ed. by N. Mollett), pp. 33–54.

Ricker W.E. (1975) *Computation and Interpretation of Biological Statistics of Fish Populations*. Department of the Environment, Fisheries, and Marine Service, Ottawa.

Robinson W.L. (1986) Individual Transferable Quotas in the Australian Southern Bluefin Tuna Fishery. In *Fishery Access Control Programs Worldwide*. (Ed. by N. Mollett), pp. 189–205.

Ruddle K. (1984) The continuity of traditional management practices: The case of Japanese coastal fisheries. In *The Traditional Management of Coastal Systems in Asia and the Pacific* (Ed. by K. Ruddle & R.E. Johannes). Unesco, Paris.

Schaefer M.B. (1954) Some aspects of the dynamics of populations important to the management of the commercial marine fisheries. *Inter-American Tropical Tuna Commission Bulletin*, **1**, 27–56.

Schaefer M.B. (1957) Some considerations of population dynamics and economics in relation to the management of marine fisheries. *Journal of the Fisheries Research Board of Canada*, **14**, 669–81.

Scott A.D. (1955) The fishery: the objectives of sole ownership. *Journal of Political Economy*, **63**, 116–24.

Smith V.L. (1975) The primitive hunter culture, pleistocene extinction, and the rise of agriculture. *Journal of Political Economy*, **83**, 727–56.

Turvey R. (1964) Optimization and suboptimization in fishery regulation. *American Economic Review*, **54**, 64–77.

Ulltang, Ø. (1980) Factors affecting the reaction of pelagic fish stocks to exploitation and requiring a new approach to assessment and management. In *The Assessment and Management of Pelagic Fish Stocks* (Ed. by A. Saville). *Conseil International pour l'Exploration de la Mer, Rapports et Procês Verbaux des Réunions*, **177**, 489–504.

United Nations (1983) *The Law of the Sea*. United Nations Convention on the Law of the Sea, with index and final act of the Third United Nations Conference on the Law of the Sea. New York.

Warming J. (1911) Om grundrente av fiskegrunde (On land rent of fishing grounds). Translated by L.G. Anderson & R. Hannesson. *Fisheries Economics Newsletter*, No. 12, Nov. 1981, Fishery Economics Research Unit, Edinburgh.

Weitzman M.L. (1974) Free access vs private ownership as alternative systems for managing common property. *Journal of Economic Theory*, **8**, 225–34.

Yamamoto T. & Short K. (1992) (eds) *International Perspectives on Fisheries Management: Proceedings of the JIFRS/IIFET/ZENGYOREN Symposium on Fisheries Management, Tokyo, 26 August–3 September 1991*. National Federation of Fisheries Cooperative Associations (ZENGYOREN) in association with Japan International Fisheries Research Society (JIFRS).

Index

Printed and bound by CPI Group (UK) Ltd, Croydon, CR0 4YY
16/09/2026
14963385-0008